Reinforced Concrete Structures

Handbook Design, Testing and Analysis

Reinforced Concrete Structures
Handbook Design, Testing and Analysis

Edward Smith

Oxford Book Company
Jaipur ▪ New Delhi

Published By

Oxford Book Company

Regd. Off. "Bony Residency", Gate No. 2
Opp. Tilak Public School, Vishweswariya Nagar,
Gopalpura Road, Jaipur - 302018, India
Ph: 0141 - 2761280, Telefax: 2761381
email - oxfordbookjpr@gmail.com
website - oxfordbookcompany.com

Delhi Office: 102, 1st Floor, Satyam House,
4327/3, Ansari Road, Darya Ganj,
New Delhi - 110002, India
Ph: 011-45652440

Distributed by
KEYA INTERNATIONAL PRESS LLP
E-mail: keyainternationalpressllp@gmail.com

First published in India, 2024

ISBN : 9789355240866

Typeset by Shivangi Computers, Jaipur

Printed and bound in India by Replika Press Pvt. Ltd.

Contents

Preface

Reinforced concrete is concrete in which reinforcing bars or other types of reinforcement have been integrated to improve one or more properties of the concrete. For many years, it has been utilized as an economical construction material in one form or another in buildings, bridges, and many other types of structures throughout the world. A large part of its worldwide appeal is that the basic constituent materials—cement, sand, aggregate, water, and reinforcing bars—are widely available and that it is possible to construct a structure using local sources of labour and materials. In addition to being readily obtainable, reinforced concrete has been universally accepted because it can be molded essentially into any shape or form, is inherently rigid, and is inherently fire-resistant. With proper protection of the reinforcement, a reinforced concrete structure can be very durable and can have a long life even under harsh climatic or environmental conditions. Reinforced concrete structures have also demonstrated that they can provide a safe haven from the potentially devastating effects of earthquakes, hurricanes, floods, and tornadoes. Based on these and other advantages, it is evident that reinforced concrete can provide viable and cost-effective solutions in a variety of applications. Concrete is a brittle, composite material that is strong in compression and weak in tension.

Cracking occurs when the concrete tensile stress in a member reaches the tensile strength due to externally applied loads, temperature changes, or shrinkage. Concrete members that do not have any type of reinforcement in them will typically fail very suddenly once the first tension cracks form because there is nothing to prevent the cracks from propagating completely through the member. In the United States and throughout the world, the design and construction of building structures is regulated by building codes. The main purpose of a building code is to protect public health, safety, and welfare. Building code provisions are founded on principles that do not unnecessarily increase construction costs; do not restrict the use of new materials, products, or methods of construction; and do not give preferential treatment to particular types or classes of materials, products, or methods of construction. Many cities, counties, and states in the United States and some international jurisdictions have adopted the International Building Code (IBC) for the design and construction of building structures.

Structural members in any structure must be designed to safely and economically support the weight of the structure and to resist all of the loads superimposed on the structure. In ordinary buildings, superimposed loads typically consist of live loads due to the inhabitants, dead loads due to items permanently attached to the building, and lateral loads due to wind or earthquakes. Some types of buildings must also be designed for extraordinary loads such as explosions or vehicular impact. A typical reinforced concrete building is made up of a variety of different reinforced concrete members. The members work together to support the applicable loads, which are transferred through load paths in the structure to the foundation members. The loads are ultimately supported by the soil or rock adjoining the foundations. Unlike other typical types of construction commonly used in building structures, such as structural steel and timber, reinforced concrete construction possesses inherent continuity. Cast-in-place reinforced concrete structures are essentially monolithic with reinforcement that extends into adjoining members. As such, reinforced concrete members are analyzed as continuous members in a statically indeterminate structure where bending moments, shear forces, and axial forces are ransferred through the joints. Understanding the behaviour and response of a reinforced concrete structure is imperative in the proper analysis, design, and detailing of the members in the structure.

The Handbook focuses in its main attractive features of the PBO method, its clarity in concepts, simplicity in mathematical formulation, ability to generate the global optimum and easy to understand.

— ***Edward Smith***

Chapter 1

Reinforced Concrete

Reinforced concrete is concrete in which reinforcement bars ("rebars"), reinforcement grids, plates or fibres have been incorporated to strengthen the concrete in tension. It was invented by French gardener Joseph Monier in 1849 and patented in 1867. The term Ferro Concrete refers only to concrete that is reinforced with iron or steel. Other materials used to reinforce concrete can be organic and inorganic fibres as well as composites in different forms. Prior to the invention of reinforcement, concrete was strong in compression, but weak in tension. Adding reinforcement crucially increases the strength in tension. The failure strain of concrete in tension is so low that the reinforcement has to hold the cracked sections together.

For a strong, ductile and durable construction the reinforcement needs to have the following properties:

- High strength
- High tensile strain
- Good bond to the concrete
- Thermal compatibility
- Durability in the concrete environment.

In most cases reinforced concrete uses steel rebars that have been inserted to add strength.

Use in Construction

Concrete is reinforced to give it extra tensile strength; without reinforcement, many concrete buildings would not have been possible.

Reinforced concrete can encompass many types of structures and components, including slabs, walls, beams, columns, foundations, frames and more.

Reinforced concrete can be classified as precast or cast in-situ concrete.

Much of the focus on reinforcing concrete is placed on floor systems. Designing and implementing the most efficient floor system is key to creating optimal building structures. Small changes in the design of a floor system can have significant impact on material costs, construction schedule, ultimate strength, operating costs, occupancy levels and end use of a building.

Behaviour of Reinforced Concrete

Materials

Concrete is a mixture of Coarse (stone or brick chips) and Fine (generally sand) aggregates with a binder material (usually Portland cement). When mixed with a small amount of water, the cement hydrates to form microscopic opaque crystal lattices encapsulating and locking the aggregate into a rigid structure. Typical concrete mixes have high resistance to compressive stresses (about 4,000 psi (28 MPa)); however, any appreciable tension (*e.g.,* due to bending) will break the microscopic rigid lattice, resulting in cracking and separation of the concrete. For this reason, typical non-reinforced concrete must be well supported to prevent the development of tension.

If a material with high strength in tension, such as steel, is placed in concrete, then the composite material, reinforced concrete, resists not only compression but also bending and other direct tensile actions. A reinforced concrete section where the concrete resists the compression and steel resists the tension can be made into almost any shape and size for the construction industry.

Key Characteristics

Three physical characteristics give reinforced concrete its special properties.

First, the coefficient of thermal expansion of concrete is similar to that of steel, eliminating large internal stresses due to differences in thermal expansion or contraction.

Second, when the cement paste within the concrete hardens this conforms to the surface details of the steel, permitting any stress to be transmitted efficiently between the different materials. Usually steel bars are roughened or corrugated to further improve the bond or cohesion between the concrete and steel.

Third, the alkaline chemical environment provided by the alkali reserve (KOH, NaOH) and the portlandite (calcium hydroxide) contained in the hardened cement paste causes a passivating film to form on the surface of the steel, making it much more resistant to corrosion than it would be in neutral or acidic conditions. When the cement paste exposed to the air and meteoric water reacts with the atmospheric CO_2, portlandite and the Calcium Silicate Hydrate (CSH) of the hardened cement paste become progressively carbonated and the high pH gradually decreases from 13.5 – 12.5 to 8.5, the pH of water in equilibrium with calcite (calcium carbonate) and the steel is no longer passivated.

As a rule of thumb, only to give an idea on orders of magnitude, steel is protected at pH above ~11 but starts to corrode below ~10 depending on steel characteristics and local physico-chemical conditions when concrete becomes carbonated. Carbonation of concrete along with chloride ingress are amongst the chief reasons for the failure of reinforcement bars in concrete.

The relative cross-sectional area of steel required for typical reinforced concrete is usually quite small and varies from 1% for most beams and slabs to 6% for some columns. Reinforcing bars are normally round in cross-section and vary in diametre. Reinforced concrete structures sometimes have provisions such as ventilated hollow cores to control their moisture & humidity.

Distribution of concrete (in spite of reinforcement) strength characteristics along the cross-section of vertical reinforced concrete elements is inhomogeneous.

Mechanism of Composite Action of Reinforcement and Concrete

The reinforcement in a RC structure, such as a steel bar, has to undergo the same strain or deformation as the surrounding concrete in order to prevent discontinuity, slip or separation of the two materials under load. Maintaining composite action requires transfer of load between the concrete and steel. The direct stress is transferred from the concrete to the bar interface so as to change the tensile stress in the reinforcing bar along its length. This load transfer is achieved by means of bond (anchorage) and is idealized as a continuous stress field that develops in the vicinity of the steel-concrete interface.

Anchorage (Bond) in Concrete: Codes of Specifications

Because the actual bond stress varies along the length of a bar anchored in a zone of tension, current international codes of specifications use the concept of development length rather than bond stress. The main requirement for safety against bond failure is to provide a sufficient extension of the length of the bar beyond the point where the steel is required to develop its yield stress and this length must be at least equal to its development length. However, if the actual available length is inadequate for full development, special anchorages must be provided, such as cogs or hooks or mechanical end plates. The same concept applies to lap splice length mentioned in the codes where splices (overlapping) provided between two adjacent bars in order to maintain the required continuity of stress in the splice zone.

Anti-corrosion Measures

In wet and cold climates, reinforced concrete for roads, bridges, parking structures and other structures that may be exposed to deicing salt may benefit from use of epoxy-coated, hot dip galvanised or stainless steel rebar, although good design and a well-chosen cement mix may provide sufficient protection for many applications. Epoxy coated rebar can easily be identified by the light green colour of its epoxy coating. Hot dip galvanized rebar may be bright or dull grey depending on length of exposure, and stainless rebar exhibits a typical white metallic sheen that is readily distinguishable from carbon steel reinforcing bar. Reference ASTM standard specifications A767 Standard Specification for Hot Dip Galvanised Reinforcing Bars, A775 Standard Specification for Epoxy Coated Steel Reinforcing Bars and A955 Standard Specification for Deformed and Plain Stainless Bars for Concrete Reinforcement. Another, cheaper way of protecting rebars is coating them with zinc phosphate. Zinc phosphate slowly reacts with calcium cations and the hydroxyl anions present in the cement pore water and forms a stable hydroxyapatite layer.

Penetrating sealants typically must be applied some time after curing. Sealants include paint, plastic foams, films and aluminium foil, felts or fabric mats sealed with tar, and layers of bentonite clay, sometimes used to seal roadbeds.

Corrosion inhibitors, such as calcium nitrite [$Ca(NO_2)_2$], can also be added to the water mix before pouring concrete. Generally, 1–2 wt. % of [$Ca(NO_2)_2$] with respect to cement weight is needed to prevent corrosion of the rebars. The nitrite anion is a mild oxidizer that oxidizes the soluble and mobile ferrous ions (Fe^{2+}) present at the surface of the corroding steel and causes it to precipitate as an insoluble ferric hydroxide ($Fe(OH)_3$). This causes the passivation of steel at the anodic oxidation sites. Nitrite is a much more active corrosion inhibitor than nitrate, a less powerful oxidizer of the divalent iron.

Reinforcement and Terminology of Beams

A beam bends under bending moment, resulting in a small curvature. At the outer face (tensile face) of the curvature the concrete experiences tensile stress, while at the inner face (compressive face) it experiences compressive stress.

A singly reinforced beam is one in which the concrete element is only reinforced near the tensile face and the reinforcement, called tension steel, is designed to resist the tension.

A doubly reinforced beam is one in which besides the tensile reinforcement the concrete element is also reinforced near the compressive face to help the concrete resist compression. The latter reinforcement is called compression steel. When the compression zone of a concrete is inadequate to resist the compressive Moment(positive moment), extra reinforcement has to be provided if the architect limits the dimensions of the section.

An under-reinforced beam is one in which the tension capacity of the tensile reinforcement is smaller than the combined compression capacity of the concrete and the compression steel (under-reinforced at tensile face). When the reinforced concrete element is subject to increasing bending moment, the tension steel yields while the concrete does not reach its ultimate failure condition.

As the tension steel yields and stretches, an "under-reinforced" concrete also yields in a ductile manner, exhibiting a large deformation and warning before its ultimate failure. In this case the yield stress of the steel governs the design.

An over-reinforced beam is one in which the tension capacity of the tension steel is greater than the combined compression capacity of the concrete and the compression steel (over-reinforced at tensile face). So the "over-reinforced concrete" beam fails by crushing of the compressive-zone concrete and before the tension zone steel yields, which does not provide any warning before failure as the failure is instantaneous.

A balanced-reinforced beam is one in which both the compressive and tensile zones reach yielding at the same imposed load on the beam, and the concrete will crush and the tensile steel will yield at the same time. This design criterion is however as risky as over-reinforced concrete, because failure is sudden as the concrete crushes at the same time of the tensile steel yields, which gives a very little warning of distress in tension failure.

Steel-reinforced concrete moment-carrying elements should normally be designed to be under-reinforced so that users of the structure will receive warning of impending collapse. The characteristic strength is the strength of a material where less than 5% of the specimen shows lower strength. The design strength or nominal strength is the strength of a material, including a material-safety factor. The value of the safety factor generally ranges from 0.75 to 0.85 in Allowable Stress Design. The ultimate limit state is the theoretical failure point with a certain probability. It is stated under factored loads and factored resistances.

Prestressed Concrete

Prestressed concrete is a technique that greatly increases loadbearing strength of concrete beams. The reinforcing steel in the bottom part of the beam, which will be subjected to tensile forces when in service, is placed in tension prior to the concrete being poured around it. Once the concrete has hardened, the tension on the reinforcing steel is released, placing a built-in compressive force on the concrete. When loads are applied, the reinforcing steel takes on more stress and the compressive force in the concrete is reduced, but does not become a tensile force. Since the concrete is always under compression, it is less subject to cracking and failure.

Common Failure Modes of Steel Reinforced Concrete

Reinforced concrete can fail due to inadequate strength, leading to mechanical failure, or due to a reduction in its durability. Corrosion and freeze/thaw cycles may damage poorly designed or constructed reinforced concrete. When rebar corrodes, the oxidation products (rust) expand and tends to flake, cracking the concrete and unbonding the rebar from the concrete. Typical mechanisms leading to durability problems are discussed below.

Mechanical Failure

Cracking of the concrete section can not be prevented; however, the size of and location of the cracks can be limited and controlled by reinforcement, placement of control joints, the curing methodology and the mix design of the concrete. Cracking defects can allow moisture to penetrate and corrode the reinforcement. This is a serviceability failure in limit state design. Cracking is normally the result of an inadequate quantity of rebar, or rebar spaced at too great a distance. The concrete then cracks either under excess loading, or due to internal effects such as early thermal shrinkage when it cures.

Ultimate failure leading to collapse can be caused by crushing of the concrete, when compressive stresses exceed its strength; by yielding or failure of the rebar, when bending or shear stresses exceed the strength of the reinforcement; or by bond failure between the concrete and the rebar.

Carbonatation

Figure : Concrete wall cracking as steel reinforcing corrodes and swells. The rust formed as the iron corrodes has a lower density than the metal, so it expands as it forms, cracking the decorative cladding off the wall, as well as damaging the structural concrete. The breakage of material from the surface is called *spalling*.

Figure : Detailed view of spalling. The apparently thin layer of concrete between the steel and the surface suggests that the design fell foul of carbonatation, or similar problems of corrosion from external exposure

When designing a concrete structure, it is normal to state the concrete cover for the rebar (the depth within the object that the rebar will be).

The minimum concrete cover is normally regulated by design or building codes. If the reinforcement is too close to the surface, early failure due to corrosion may occur.

The concrete cover depth can be measured with a cover metre. However, carbonatated concrete only becomes a durability problem when there is also sufficient moisture and oxygen to cause electro-potential corrosion of the reinforcing steel.

One method of testing a structure for carbonatation is to drill a fresh hole in the surface and then treat the cut surface with phenolphthalein indicator solution. This solution will turn [pink] when in contact with alkaline concrete, making it possible to see the depth of carbonatation. An existing hole is no good because the exposed surface will already be carbonatated.

Chlorides

Figure : The Paulins Kill Viaduct, Hainesburg, New Jersey, is 115 feet (35 m) tall and 1,100 feet (335 m) long, and was heralded as the largest reinforced concrete structure in the world when it was completed in 1910 as part of the Lackawanna Cut-Off rail line project. The Lackawanna Railroad was a pioneer in the use of reinforced concrete.

Chlorides, including sodium chloride, can promote the corrosion of embedded steel rebar if present in sufficienty high concentration. Chloride anions induce both localized corrosion (pitting corrosion) and generalised corrosion of steel reinforcements. For this reason, one should only use fresh raw water or potable water for mixing concrete, ensure that the coarse and fine aggregates do not contain chlorides, and not use admixtures that contain chlorides.

It was once common for calcium chloride to be used as an admixture to promote rapid set-up of the concrete. It was also mistakenly believed that it would prevent freezing. However, this practice has fallen into disfavour once the deleterious effects of chlorides became known. It should be avoided when ever possible.

The use of de-icing salts on roadways, used to reduce the freezing point of water, is probably one of the primary causes of premature failure of reinforced or prestressed concrete bridge decks, roadways, and parking garages. The use of epoxy-coated reinforcing bars and the application of cathodic protection has mitigated this problem to some extent. Also FRP rebars are known to be less susceptible to chlorides. Properly designed concrete mixtures that have been allowed to cure properly are effectively impervious to the effects of deicers.

Another important source of chloride ions is from sea water. Sea water contains by weight approximately 3.5 wt.% salts. These salts include sodium chloride, magnesium sulfate, calcium sulfate, and bicarbonates. In water these salts dissociate in free ions (Na^+, Mg^{2+}, Cl^-, SO_4^{2-}, HCO_3^-) and migrate with the water into the capillaries of the concrete. Chloride ions are particularly aggressive for the corrosion of the carbon steel reinforcement bars and make up about 50% of these ions.

In the 1960's and 1970's it was also relatively common for Magnesite, a chloride rich carbonate mineral, to be used as a floor-topping material. This was done principally as a levelling and sound attenuating layer. However it is now known that when these materials came into contact with moisture it produced a weak solution of hydrochloric acid due to the presence of chlorides in the magnesite. Over a period of time (typically decades) the solution caused corrosion of the embedded steel rebars. This was most commonly found in wet areas or areas repeatedly exposed to moisture.

Alkali Silica Reaction

This a reaction of amorphous silica (chalcedony, chert, siliceous limestone) sometimes present in the aggregates with the hydroxyl ions (OH^-) from the cement pore solution. Poorly crystallized silica (SiO_2) dissolves and dissociates at high pH (12.5 - 13.5) in alkaline water. The soluble dissociated silicic acid reacts in the porewater with the calcium hydroxide (portlandite) present in the cement paste to form an expansive calcium silicate hydrate (CSH). The *alkali silica reaction (ASR)*, causes localised swelling responsible of tensile stress and cracking. The conditions required for alkali silica reaction are threefold: (1) aggregate containing an alkali-reactive constituent (amorphous silica), (2) sufficient availability of hydroxyl ions (OH^-), and (3) sufficient moisture, above 75 % relative humidity (RH) within the concrete. This phenomenon is sometimes popularly referred to as "concrete cancer".

Conversion of High Alumina Cement

Resistant to weak acids and especially sulfates, this cement cures quickly and reaches very high durability and strength. It was greatly used after World War II for making precast concrete objects. However, it can lose strength with heat or time (conversion), especially when not properly cured. With the collapse of three roofs made of prestressed concrete beams using high alumina cement, this cement was banned in the UK in 1976. Subsequent inquiries into the matter showed that the beams were improperly manufactured, but the ban remained.

Sulphates

Sulphates (SO_4) in the soil or in groundwater, in sufficient concentration, can react with the Portland cement in concrete causing the formation of expansive products, e.g. ettringite or thaumasite, which can lead to early failure of the structure. The most typical attack of this type is on concrete slabs and foundation walls at grade where the sulfate ion, via alternate wetting and drying, can increase in concentration. As the concentration increases, the attack on the Portland cement can begin.

For buried structures such as pipe, this type of attack is much rarer especially in the Eastern half of the United States. The sulfate ion concentration increases much slower in the soil mass and is especially dependent upon the initial amount of sulfates in the native soil. The chemical analysis of soil borings should be done during the design phase of any project involving concrete in contact with the native soil to check for the presence of sulfates. If the concentrations are found to be aggressive, various protective coatings can be used. Also, in the US ASTM C150 Type 5 Portland cement can be used in the mix. This type of cement is designed to be particularly resistant to a sulfate attack.

Steel Plate Construction

In steel plate construction, stringers join parallel steel plates. The plate assemblies are fabricated off site, and welded together on-site to form steel walls connected by stringers. The walls become the form into which concrete is poured. Steel plate construction speeds reinforced concrete construction by cutting out the time consuming on-site manual steps of tying rebar and building forms. The method has excellent strength because the steel is on the outside, where tensile forces are often greatest.

Faux Bois

Figure : Torii Gate by Dionicio Rodriguez at Brackenridge Park, San Antonio, Texas.

Figure : Torii Gate by Dionicio Rodriguez at Brackenridge Park, San Antonio, Texas

Faux bois (from the French for *false wood*) refers to the artistic imitation of wood or wood grains in various media. The craft has roots in the Renaissance with trompe-l'œil. It was probably first crafted with concrete using a steel armature by the inventor of ferrocement, Joseph Monier. In 1875, Monier created the first bridge of reinforced concrete at Chazelet, France. It was sculpted to resemble timbers and logs.

Ferrocement faux bois uses a combination of concrete, mortar and grout applied to a steel frame or armature to sculpt life-like representations of wooden objects. Final sculpting can be done while the mixture is wet, in a putty state, or slightly stiff. Techniques vary among artisans. Most popular in the late 19th century through the 1940s, ferrocement faux bois has largely disappeared with the passing of those most expert in its practice. What few objects remain from that peak period (mostly in the form of garden art, such as planters and birdbaths) are now highly prized by collectors.

In Mexico, this style was known as "El Trabajo Rústico" (*The Rustic Work*). One highly regarded artist who worked in this style was Dionicio Rodriguez, a Mexican who relocated to Texas in the early 1920s. Although not widely known, his large-scale faux bois installations have been listed on the National Register of Historic Places. Dionicio's great-nephew is one of the handful of artists still creating Faux Bois today.

Protection of Exposed Concrete

Protection of exposed concrete is necessary to prolong its service life. A 120 year design life for concrete infrastructure has become increasingly common. Without early preventive maintenance this design life target may be optimistic. The design service life of reinforced concrete structures often is not reached because of early deterioration and damage. The service life of some modern concrete structures is 20 to 30 years at most. Even good quality engineering concrete is made up of countless invisible interconnecting capillary pores. The concrete acts like a "hard sponge" which absorbs damaging liquids such as water and aggressive water-borne salts. The strength of any concrete lessens as a result of static and dynamic loading, chloride ion ingress (from sea water) and the application of deicing salts. This leaves the concrete vulnerable to corrosion of the embedded steel. Concrete can also be damaged by sulfate attack, alkali silica reaction, carbonation, temperature change, abrasion, biological attack, salt crystallization, efflorescence and freeze-thaw attack. Many of these processes are surface water driven and can be mitigated by early preventive maintenance with alkyl alkoxy silane impregnation.

Capillary Suction

Water and salts are drawn into the capillaries of the concrete via capillary suction and wicking, i.e. water moving from the saturated zone to the dry zone. The "war zones" for marine concrete are those areas that are subjected to the constant wet-dry cycles of wave action and wind. This causes oxygen, water and chloride ions to become plentiful. The corners of concrete in these areas are particularly vulnerable as they are attacked from two directions.

Capillary action is caused by surface tension and by the relative value of the adhesion between the water and the concrete to the cohesion of the water. The action of surface tension is to cause the water to rise within a small capillary that is partially immersed in the water. The water can rise to its maximum height occurs when the contact angle is zero. The smaller the pore radius the higher the water can rise. This is the same mechanism that gets water from the root of a tree to the tip of its highest leaf.

Chloride Ion

Sea water contains approximately 3.5 wt. % salts by weight including; sodium chloride, magnesium sulphate, calcium sulphate and bicarbonates. In water these salts dissociate and migrate with the water into the capillaries of the concrete. Chloride ions are particularly aggressive for the corrosion of steel reinforcement bars and make up about 50 % of these ions.

Chloride ions are easily incorporated in the crystal lattice of hydrated cement Afm phases where they form Friedel's salt. These salts can only penetrate into the concrete when dissolved in water. If the water is stopped from penetrating the concrete so are the salts it contains.

Saline water from the sea or de-icing salts (e.g. calcium chloride) provides a strong electrolyte to help drive the electro-potential corrosion of reinforcing steel. Local micro-environmental differences along the length of the reinforcement are enough to set up a potential difference and initiate corrosion. Chloride ions act a catalyst pulling the iron ions into solution to form rust in the presence of water and oxygen, resulting in the pitting corrosion of the steel. Rust can cause iron to expand 3-7 times its original size and this causes high tensile stress in surrounding concrete.

If the stress is great enough the concrete will crack, spall and may delaminate the concrete cover. De-icing salts such as calcium chloride and sodium chloride are added to road surfaces in winter to lower the freezing point of any surface water (over 50 million tons of de-icing salts are used each year on US roads). When it does freeze, unlike most liquids, water expands by 9 %. The cyclic freeze thaw action causes great pressures e.g. 200 MPa to build up in the pores causing micro-cracking and scaling.

Sea water and some ground water contain many aggressive salts such as sodium chloride, calcium chloride, magnesium sulphate, sodium sulphate, calcium sulphate and bicarbonates. When these salts dry in the surface pores the resulting crystallization causes reduced cohesion of the cement paste, softens the concrete and reduces its strength.

Alkali Silica Reaction

When the alkalis in cement react with amorphous (i.e. non crystalline) silica aggregate in the presence of water then a disruptive swelling reaction occurs in the cement paste. When this happens to concrete railway sleepers the rails may end up out of alignment and cause a derailment.

Preventive Treatment

Relative to the capillary pore size of concrete the alkyl alkoxy silane molecule is very small and can penetrate deeply. There are a number of silanes available and these vary with the alkyl part of the molecule to suit the application. For example some alkyl groups can only repel water but not oil and others can repel both water and oil. The preferred alkoxy group is ethoxy. The reason for this is because the ethoxy is a relatively slow reacting group allowing the silane to penetrate deeply, even displacing water that may be present, before reacting. Also, this group over time reacts with water in the pores or air to form ethanol which is a far safer and environmentally sound by-product than a methoxy group which would form methanol.

The result is a permanent connection with the capillary pore walls that repels liquids. The pores remain unblocked and allow the passage of water vapour, which being a gas has no surface tension. Allowing water vapour to escape has the effect of increasing the resistivity of the concrete and thus slowing down the rate of electro-potential corrosion of the reinforcing steel. Concrete can be silane-treated from the low tide level and above. It may take several treatments some weeks apart to obtain a good depth of impregnation between the low and high tide levels. Multiple applications over time allow the concrete to dry and allow for the silane to penetrate. Even though concrete with a pore diametre of say 2×10^{-6} m can temporarily resist over 4 m head of water pressure (P = -4x surface tension × Cos (contact angle) /pore diametre) i.e. only exposed surface should be treated . Below the low tide mark the available oxygen levels drops off steeply. At 15 °C at sea level there is only 7 mg of oxygen dissolved per litre of sea water. Air contains 250 mg of oxygen per litre, i.e. 36 time more oxygen available than in water.

When the capillaries are treated with silane the contact angle typically become 110 °. In the equation Cos (110 °) is a negative number and this describes the result of the treatment i.e. a repulsion of the water and salts. Only the pores are treated and there is no film or coating on the surface and so there is little or no change in its frictional properties or appearance.

Low Volatile Organic Compounds (VOCs) impregnating silane water-based creams are now available. VOCs include any volatile compound of carbon that participates in atmospheric photochemical reactions. In water repellents and related materials, a VOC is typically a formulation ingredient that will evaporate (volatilize) under normal use and is expressed as grams per litre. On July 1, 2006, the South Coast Air Quality Management District (SCAQMD) in California implemented the most stringent VOC requirements in the world. It is reasonable to expect that the rules we see in place in California today will influence a large portion of the world in the next few years. For SCAQMD, [waterproof | [waterproofing]] concrete sealers need to have VOC content below (excluding water and exempt compounds) 100 grams per Litre. The VOC content does not change regardless of how much water or exempt solvent is added to the sealer.

Silane cream can be applied in one coat using a foam roller or low pressure spray unit. The typical application rate is 1 litre per 3 to 4 sq.m. depending on the surface absorption and depth of penetration required. It needs to be protected from moisture for a minimum of 12 hours. The 21 day reduction in 15 % NaCl brine water uptake in accordance NCHRP Report 244 method is up to 70 % for liquid silane and 64 % for cream.

The cream combines the best aspects of both water-based and solvent-based technologies. The cream sits on the surface and once the water evaporates the solvents take over to penetrate deeply into the surface. Unlike a typical water-based coating the cream can even pass through a previously impregnated (i.e. water repelling surface). The cream is applied in one coat and allows the applicator better control particularly in windy conditions and on vertical and overhead surfaces.

If not used correctly, solvent-based silanes pose significant health and safety problems with potential risk of skin and eye irritation and damage to vegetation and aquatic environments. There is little risk in using cream-based products.

Impregnation undertaken in the UK is known to be effective for at least 15 years, provided it is applied correctly. Longer service life is anticipated.

Service Life Prediction

With construction standards in the United States and several other countries turning towards performance-based requirements, concrete service life prediction has grown in importance. Service life modelling software (such as STADIUM) is now recommended by the US Navy for new waterfront construction.

Case Study

The quay-wall of a new container terminal at Zeebrugge Harbor, Belgium, was protected against chloride ingress by means of a water-repellent agent immediately after construction in 1993. An alkyl triethoxy silane was used pre-evaluated in a preliminary research program. To judge the in site effectiveness of the hydrophobic agent as a water repellent treatment, three subsequent in site surveys were conducted in 1996, 1998 and recently in 2005. Based on the cores drilled, the chloride profiles are determined as a function of time, both in a non-treated and treated location. Because of the long-term data sequence, the long-term effectiveness of the treatment can be assessed in an objective way. Predicted service life represents the 50 % probability of chloride ions to reaching reinforcing bars at a depth of 120 mm with a sufficient concentration of 0.7 % chloride ion by weight of cement to start the corrosion process i.e. untreated 16.5 years, treated 107 years.

Depth of Impregnation

Amongst the most decisive parameters determining the effectiveness of a water repellent treatment is the penetration depth. A good depth of penetration not only protects the treatment from weathering and traffic but also sets up a hydrophobic barrier. The depth of penetration is tested by breaking open the treated specimen and spraying the fractured surface with water. The depth of the dry zone is taken as the effective depth of impregnation.

This barrier stops the passage of liquid water and salts into or out of the concrete. Depth of penetration is also a good quality control check. A 50 mm diametre 40 mm deep button core is recommended to be randomly taken for every 300 sq.m. of concrete treated. This can be compared with a known sample to ensure that the contractor is applying the correct quantity of silane. It should be greater than or equal to 5 mm. It should be noted that existing treated surfaces can easily absorb more applications and the new material will penetrate deeper levels in the concrete. This is enhanced by waiting for each application to fully dry.

Alberta Transportation and Utilities carried out a series of five day water immersion tests to measure the reduction in water uptake on concrete samples treated with common sealers i.e. epoxy, acrylic, siloxane and silane. The water uptake was also measured after sandblasting to abrade (total of 3 % removed by weight) the surface as a means of simulated accelerated weathering. Only the silanes kept on performing after significant surface abrasion. It is estimated that a highway concrete wearing surface will lose 1 mm of concrete every 7 years.

Theoretical Break Through Pressure

The anti-capillary force caused by the silane treatment can be overcome by applying a sufficiently high force such as extreme wind driven rain pressure or by hydrostatic pressure on the liquid trying to enter the pore. The amount of force needed to push the water into the concrete is proportional to the imparted critical surface tension of the substrate and the pore's diametre.

$$\text{Break through pressure} = \frac{-4\gamma\cos\theta}{d}$$

where:

γ : the liquid-air surface tension (energy/surface area)

θ : the contact angle (angle)

d: diametre of pore (length)

For a concrete, using SI units:

$\gamma = 0.072$ J/m^2 at 20 °C

$\theta = 107$ ° (angle)

$d = 2.1 \times 10^{-6}$ m

Breakthrough pressure is 40,100 N/m^2 or 4.092 metres (i.e. 40100/ 101325 × 10.34 = 4.092 m) pressure head of water. By comparison 50 mm pressure head is the same pressure from 104 km per hour wind driven rain.

If the pore size increases, the head pressure resistance will decrease. Water vapour does not have a surface tension, it can pass freely through a substrate made water repellent with the silane treatment.

Conclusion

Many of the processes that deteriorate the strength of engineering concrete are surface-water driven and can be mitigated by early preventive maintenance with an alkyl alkoxy impregnating silane. The silane is used to line capillary pores and make them hydrophobic stopping capillary suction and wicking action.

These products are non-film forming; able to greatly reduce water uptake; an excellent chloride ion screen; highly water vapour permeable; deeply penetrating; very alkali resistant; do not change the appearance or frictional property of the surface and can seal hairline cracks.

Cover Metre

A cover metre is an instrument to locate rebars and measure the exact concrete cover. Rebar detectors are less sophisticated devices that can only locate metallic objects below the surface. Due to the cost-effective design, the pulse-induction method is one of the most commonly used solutions.

Method

The pulse-induction method is based on electromagnetic pulse induction technology to detect rebars. Coils in the probe are periodically charged by current pulses and thus generate a magnetic field. On the surface of any electrically conductive material which is in the magnetic field eddy currents are produced. They induce a magnetic field in opposite directions. The resulting change in voltage can be utilised for the measurement. Rebars that are closer to the probe or of larger size produce a stronger magnetic field.

Modern rebar detectors use different coil arrangements to generate several magnetic fields. Advanced signal processing supports not only the localization of rebars but also the determination of the cover and the estimation of the bar diametre. This method is unaffected by all non conductive materials such as concrete, wood, plastics, bricks, etc. However any kind of conductive materials within the magnetic field will have an influence on the measurement.

Advantages of the pulse induction method:

- high accuracy
- not influenced by moisture and heterogeneities of the concrete
- unaffected by environmental influences
- low costs

Disadvantage of the pulse induction method:

- Limited detection range
- Minimum bar spacing depends on cover depths.

Standards

- BS1881:204 Testing concrete. Recommendations on the use of electromagnetic covermetres
- DGZfP:B2: Guideline “für Bewehrungsnachweis und Überdeckungsmessung bei Stahl- und Spannbeton”
- DIN 1045: Guideline Concrete, reinforced and prestressed concrete structures
- ACI Concrete Practices Non Destructive testing 228.2R-2.51: Covermetres.

Application

Early diagnosis and analysis of seemingly healthy concrete cover and reinforcement status allows pre-emptive corrosion control measures to reduce unwanted risks to structural safety. Bundesanstalt für Materialforschung und -prüfung (Federal Institute for Materials Research and Testing, Germany) has developed a sensor equipped robotic system to accelerate the collection of several criteria used for diagnostics. Besides ultrasonic, ground-penetrating radar, concrete resistance, potential field, the eddy current method implemented in the Profometre 5 was used to measure the concrete cover.

Rebar

A rebar (short for reinforcing bar), also known as reinforcing steel, reinforcement steel, rerod, or a deformed bar, is a common steel bar, and is commonly used as a tensioning device in reinforced concrete and reinforced masonry structures holding the concrete in compression. It is usually formed from carbon steel, and is given ridges for better mechanical anchoring into the concrete. In Australia, it is colloquially known as reo.

History

Rebars were known in construction well before the era of the modern reinforced concrete. Some 150 years before its invention rebars were used to form the carcass of the Leaning Tower of Nevyansk in Russia, built on the orders of the industrialist Akinfiy Demidov. The purpose of such construction is one of the many mysteries of the tower. The cast iron used for rebars was of very high quality, and there is no corrosion on them up to this day. The carcass of the tower was connected to its cast iron tented roof, crowned with the first lightning rod in the Western world. This lightning rod was grounded through the carcass, though it is not clear whether the effect was intentional.

Use in Concrete and Masonry

Concrete is a material that is very strong in compression, but relatively weak in tension. To compensate for this imbalance in concrete's behaviour, rebar is cast into it to carry the tensile loads.

Masonry structures and the mortar holding them together have similar properties to concrete and also have a limited ability to carry tensile loads. Some standard masonry units like blocks and bricks are made with strategically placed voids to accommodate rebar, which is then secured in place with grout. This combination is known as reinforced masonry. While any material with sufficient tensile strength could conceivably be used to reinforce concrete, steel and concrete have similar coefficients of thermal expansion: a concrete structural member reinforced with steel will experience minimal stress as a result of differential expansions of the two interconnected materials caused by temperature changes.

Physical Characteristics

Steel has an expansion coefficient nearly equal to that of modern concrete. If this were not so, it would cause problems through additional longitudinal and perpendicular stresses at temperatures different than the temperature of the setting. Although rebar has ribs that bind it mechanically to the concrete, it can still be pulled out of the concrete under high stresses, an occurrence that often precedes a larger-scale collapse of the structure. To prevent such a failure, rebar is either deeply embedded into adjacent structural members (40-60 times the diametre), or bent and hooked at the ends to lock it around the concrete and other rebar. This first approach increases the friction locking the bar into place, while the second makes use of the high compressive strength of concrete. Common rebar is made of unfinished tempered steel, making it susceptible to rusting. Normally the concrete cover is able to provide a pH value higher than 12 avoiding the corrosion reaction. Too little concrete cover can compromise this guard through carbonation from the surface. Too much concrete cover can cause bigger crack widths which also compromises the local guard. As rust takes up greater volume than the steel from which it was formed, it causes severe internal pressure on the surrounding concrete, leading to cracking, spalling, and ultimately, structural failure. This is a particular problem where the concrete is exposed to salt water, as in bridges built in areas where salt is applied to roadways in winter, or in marine applications. Epoxy-coated, galvanized or stainless steel rebars may be employed in these situations at greater initial expense, but significantly lower expense over the service life of the project. Special care must be taken during the installation of epoxy-coated rebar, because even small cracks and failures in the coating can lead to intensified local chemical reactions not visible at the surface. Fibre-reinforced polymer rebar is now also being used in high-corrosion environments. It is available in many forms, from spirals for reinforcing columns, to the common rod, to meshes and many other forms. Most commercially available rebars are made from unidirectional glassfibre reinforced thermoset resins.

Sizes and Grades

U.S. Sizes

Imperial bar designations represent the bar diametre in fractions of [! inch, such that #8 = 8D 8 inch = 1 inch diametre. Area = (bar size/9)^2 such that area of #8 = (8/9)^2 = 0.79 in^2. This applies to #8 bars and smaller. Larger bars have a slightly larger diametre than the one computed using the [! inch convention.

Imperial Bar Size	*"Soft" Metric Size*	*Weight per unit length (lb/ft)*	*Mass per unit length (kg/m)*	*Nominal Diameter (in)*	*Nominal Diameter (mm)*	*Nominal Area (in^2)*	*Nominal Area (mm^2)*
#3	#10	0.376	0.561	0.375 = ?	9.525	0.11	71
#4	#13	0.668	0.996	0.500 = ½	12.7	0.20	129
#5	#16	1.043	1.556	0.625 = ?	15.875	0.31	200
#6	#19	1.502	2.24	0.750 = ¾	19.05	0.44	284
#7	#22	2.044	3.049	0.875 = ?	22.225	0.60	387
#8	#25	2.670	3.982	1.000	25.4	0.79	509
#9	#29	3.400	5.071	1.128	28.65	1.00	645
#10	#32	4.303	6.418	1.270	32.26	1.27	819
#11	#36	5.313	7.924	1.410	35.81	1.56	1006
#12	#40	6.424	9.619	1.50	38.1	1.76	1140
#14	#43	7.650	11.41	1.693	43	2.25	1452
#18	#57	13.60	20.284	2.257	57.33	4.00	2581

Canadian Sizes

Metric bar designations represent the nominal bar diametre in millimetres, rounded to the nearest 5 mm.

Metric Bar Size	*Mass per unit length (kg/m)*	*Nominal Diametre (mm)*	*Cross-Sectional Area (mm^2)*
10M	0.785	11.3	100
15M	1.570	16.0	200
20M	2.355	19.5	300
25M	3.925	25.2	500
30M	5.495	29.9	700

35M	7.850	35.7	1000
45M	11.775	43.7	1500
55M	19.625	56.4	2500

European Sizes

Metric bar designations represent the nominal bar diametre in millimetres. Bars in Europe will be specified to comply with the standard EN 10080 (awaiting introduction as of early 2007), although various national standards still remain in force (e.g. BS 4449 in the United Kingdom).

Metric Bar Size	*Mass per unit length (kg/m)*	*Nominal Diametre (mm)*	*Cross-Sectional Area (mm²)*
6,0	0.222	6	28.3
8,0	0.395	8	50.3
10,0	0.617	10	78.5
12,0	0.888	12	113
14,0	1.21	14	154
16,0	1.579	16	201
20,0	2.467	20	314
25,0	3.855	25	491
28,0	4.83	28	616
32,0	6.316	32	804
40,0	9.868	40	1257
50,0	15.413	50	1963

Grades

Rebar is available in different grades and specifications that vary in yield strength, ultimate tensile strength, chemical composition, and percentage of elongation.

The grade designation is equal to the minimum yield strength of the bar in ksi (1000 psi) for example grade 60 rebar has a minimum yield strength of 60 ksi. Rebar is typically manufactured in grades 40, 60, and 75.

Common ASTM specification are:

- ASTM A82: Specification for Plain Steel Wire for Concrete Reinforcement
- ASTM A184/A184M: Specification for Fabricated Deformed Steel Bar Mats for Concrete Reinforcement

- ASTM A185: Specification for Welded Plain Steel Wire Fabric for Concrete Reinforcement
- ASTM A496: Specification for Deformed Steel Wire for Concrete Reinforcement
- ASTM A497: Specification for Welded Deformed Steel Wire Fabric for Concrete Reinforcement
- ASTM A615/A615M: Deformed and plain carbon-steel bars for concrete reinforcement
- ASTM A616/A616M: Specification for Rail-Steel Deformed and Plain Bars for Concrete Reinforcement
- ASTM A617/A617M: Specification for Axle-Steel Deformed and Plain Bars for Concrete Reinforcement
- ASTM A706/A706M: Low-alloy steel deformed and plain bars for concrete reinforcement
- ASTM A767/A767M: Specification for Zinc-Coated(Galvanized) Steel Bars for Concrete Reinforcement
- ASTM A775/A775M: Specification for Epoxy-Coated Reinforcing Steel Bars
- ASTM A934/A934M: Specification for Epoxy-Coated Prefabricated Steel Reinforcing Bars
- ASTM A955: Deformed and plain stainless-steel bars for concrete reinforcement
- ASTM A996: Rail-steel and axle-steel deformed bars for concrete reinforcement

ASTM marking designations are:

- 'S' billet A615
- 'I' rail A616
- 'IR' Rail Meeting Supplementary Requirements S1 A616
- 'A' Axle A617
- 'W' Low-alloy — A706

Historically in Europe, rebar is composed of mild steel material with a yield strength of approximately 250 N/mm^2. Modern rebar is composed of high-yield steel, with a yield strength more typically 500 N/mm^2. Rebar can be supplied with various grades of ductility, with the more ductile steel capable of absorbing considerably greater energy when deformed - this can be of use in design to resist the forces from earthquakes for example.

Placing Rebar

Rebar cages are fabricated either on or off the project site commonly with the help of hydraulic benders and shears, however for small or custom work a tool known as a Hickey - or hand rebar bender, is sufficient. The rebars are placed by *rodbusters* or concrete reinforcing ironworkers with bar supports separating the rebar from the concrete forms to establish concrete cover and ensure that proper embedment is achieved. The rebars in the cages are connected either by welding, tying steel wire, or with mechanical connections. For epoxy coated or galvanised rebars only the latter is possible.

Welding

The American Welding Society (AWS) D 1.4 sets out the practices for welding rebar in the U.S. Without special consideration the only rebar that is ready to weld is *W grade* (Low-alloy — A706). Rebar that is not produced to the ASTM A706 specification is generally not suitable for welding without calculating the "carbon-equivalent". Material with a carbon-equivalent of less than 0.55 can be welded. (AWS D1.4). ASTM A 616 & ASTM A 617 reinforcing are re-rolled rail steel & re-rolled rail axle steel with uncontrolled chemistry, phosphorus & carbon content. These materials are not common.

Rebar cages are normally tied together with wire, although welding of cages has been the norm in Europe for many years, and is becoming more common in the US. High strength steels.

Mechanical Connections

Also known as "mechanical couplers" or "mechanical splices", mechanical connections are used to connect reinforcing bars together. Mechanical couplers are an effective means to reduce rebar congestion in highly reinforced areas for cast-in-place concrete construction. These couplers are also used in precast concrete construction at the joints between members. The structural performance criteria for mechanical connections varies considerably between different countries, codes, and industries. As a minimum requirement, codes typically specify that the rebar to splice connection meets or exceeds 125% of the specified tensile strength of the rebar. More stringent criteria also requires the development of the specified ultimate strength of the rebar. As an example, ACI 318 specifies either Type 1 (125% F_y) or Type 2 (125% F_y and 100% F_u) performance criteria.

For concrete structures designed with ductility in mind, it is recommended that the mechanical connections are also capable of failing in a ductile manner, typically known in the reinforcing steel industry as achieving "bar-break". As an example, Caltrans specifies a required mode of failure (i.e., "necking of the bar").

Safety

To prevent workers and / or pedestrians from accidentally impaling themselves, the protruding ends of steel rebar are often bent over or covered with special steel-reinforced plastic "plate" caps. "Mushroom" caps may provide protection from scratches and other minor injuries, but provide little to no protection from impalement.

Designations

For clarity, reinforcement is usually tabulated in a "reinforcement schedule" on construction drawings. This eliminates ambiguity in the various notations used in different parts of the world. The following list provides examples of the different notations used in the architectural, engineering, and construction industry.

New Zealand

Designation	*Explanation*
HD-16-300, T&B, EW	High strength (500 MPa) 16 mm diametre rebars spaced at 300 mm centres (centre-to-centre distance) on both the top and bottom face and in each way as well (i.e., longitudinal and transverse).
3-D12	Three mild strength (300 MPa) 12 mm diametre rebars
R8 Stirrups @ 225 MAX	*D* grade (300 MPa) smooth bar stirrups, spaced at 225 mm centres. By default in New Zealand practice all stirrups are normally interpreted as being full, closed, loops. This is a detailing requirement for concrete ductility in seismic zones; If a single strand of stirrup with a hook at each end was required, this would typically be both specified and illustrated.

United States

Designation	*Explanation*
#4 @ 12 OC, T&B, EW	Number 4 rebars spaced 12 inches on centre (centre-to-centre distance) on both the top and bottom faces and in each way as well, i.e. longitudinal and transverse.
(3) #4	Three number 4 rebars (usually used when the rebar perpendicular to the detail)

#3 ties @ 9 OC, (2) per set	Number 3 rebars used as stirrups, spaced at 9 inches on centre. Each set consists of two ties, which is usually illustrated.
#7 @ 12" EW, EF	Number 7 rebar spaced 12 inches apart, placed in each direction (each way) and on each face.

Carbon Grid

Carbon Grid FRP Structures for use in reinforcing concrete:

Technology: Carbon fibre FRP Grids are structural reinforcement materials that improve the performance of concrete structures such as insulated wall panels, architectural panels, double tee parking garage tee beams, concrete countertops and other products. The carbon grids can be used in place of welded wire mesh in most structures and is imbedded the same way as a welded wire mesh. Carbon fibre's high strength, high modulus, resistance to creep and excellent fatigue properties allow the carbon grid to provide excellent crack and structural reinforcement. Ideally the carbon tows (yarns) are constructed on top of each other in a superimposed design those aides with fibre alignment and toughness of the finished grid.

Applications with Concrete in General: Concrete is inherently strong in compression and weak in tension. To address this issue, concrete is often reinforced with steel in the form of rebar, welded wire mesh or stressing strands (Prestressed or post-tensioned). In theory, the concrete has to crack when loaded in tension for the steel to begin to share the load in the steel reinforced concrete composite. The concrete helps protect the steel by providing an alkaline environment (ph=13 in many cases) to retard corrosion of the steel. To properly protect the steel the concrete needs to cover the steel by a minimum thickness, not have large cracks and not have it's chemistry altered by environmental factors like chloride attack from deicing salts, carbonation, etc.. ACI and PCI (American Concrete Institute, Precast/Prestressed Concrete Institute) codes specify minimum cover thicknesses depending on the application of the structure. In practice, steel often corrodes due to improper placement of steel reinforcement, moisture drive through cracks, poor workmanship, environmental effects such as deicing salts or costal environments and a myriad of other causes. Due to the inherent properties of steel reinforced concrete, many structures are expensive, heavy and costly to maintain. The repair of these structures is a multi-billion dollar business worldwide according to industry sources (ICRI International Concrete Repair Institute).

Applications with Concrete Panels: Multi layer concrete structures (like insulated walls panels) use a sandwich technology that use insulation layers placed in between two or more layers of concrete. Often these concrete layers are tied together by sections of solid concrete, metal tie structures or FRP pin structures. Often, the design of the wall panel is limited by lack of effective insulation in the tie areas or poor structural performance of the tie design or solid section. These limitations cause structures to be weaker, heavier and more costly both in manufacturing and ownership.

Benefits: Carbon grids provide a corrosion resistant reinforcement that has a higher modulus of elasticity than steel that is imbedded inside concrete (similar to conventional steel mesh). This allows for designs that don't require as much concrete cover, so concrete structures can be lighter and much more durable than before. For instance, using carbon fibre grid in insulated wall panels can provide a stronger, lighter, more thermally efficient panel than is practical with conventional designs. Due to the lighter weight less energy is required to provide the necessary concrete and ship products adding an environmental incentive to using this technology.

FRP fabrics that are used as external reinforcements are impregnated with polymer on the jobsite and have to be bonded to a surface of a concrete structure. In contrast, carbon grids provide an internal reinforcement and have fibres that are impregnated with a polymer (typically epoxy) in a factory rather than on the job site, saving time and insuring consistent quality. Also, the open spaces of the grid provide a mechanical bond to transfer load within the concrete structure.

Additional benefits of the carbon grid structure are that it is light, easy to handle, non-magnetic and easier to cut than steel. These properties allow for use where steel is impractical or cumbersome.

History: Carbon grids have been manufactured by Chomarat North America (formerly TechFab LLC) since 1998 and have found commercial uses in concrete countertops, shotcrete, ornamental concrete, caststone products, precast insulated wall panels, double tee beams and a myriad of other concrete products. These materials have also been used in numerous repair projects including the Naumburg Band shell in New York City. A variety of product strengths and apertures are offered depending on the application.

Limitations: The main limitation of the carbon grid is the additional cost of carbon as compared to mild carbon steel (with rising energy costs this gap has narrowed). In addition, since carbon is a linear elastic material (without a yield point), reinforced concrete structures have to be designed to account for this. Carbon fibre grids can be used in design by following the American Concrete Institute - ACI 440 methodology similar to FRPrebar. Typically, the structure will also use steel in an appropriate location or be over designed so that the structure fails by crushing the concrete in tension providing ductility.

Concrete Cover

Concrete cover, in reinforced concrete, is the least distance between the surface of embedded reinforcement and the outer surface of the concrete (ACI 130). The concrete cover depth can be measured with a cover metre.

Purpose of Provision of Concrete Cover

The concrete cover must have a minimum thickness for three main reasons:

- to protect the steel reinforcement bars (rebars) from environmental effects to prevent their corrosion;
- to protect the reinforcement bars from fire, and;
- to give reinforcing bars sufficient embedding to enable them to be stressed without slipping.

The premature failure of corroded steel reinforcements and the expansion of the iron corrosion products around the rebars are amongst the main causes of the concrete degradation. The carbon steel of rebars is protected from oxidation by atmospheric oxygen by the high pH of concrete interstitial water. Iron bar surface is passivated as long as the pH value is higher than 10.5. Fresh cement water has a pH of about 13.5 while evolved cement water pH ~ 12.5 is controlled by the dissolution of calcium hydroxide (portlandite).

Carbon dioxide present in the air slowly diffuses through the concrete cover over the rebar and progressively reacts with the alkaline hydroxides (KOH, NaOH) and with calcium hydroxide leading to the carbonation of the hydrated cement paste. As a result, the pH of the cement drops and when its value is below 10.5 – 9.5, steel surface is no longer passivated and starts to corrode. A sufficient thickness of concrete cover is thus required in order to slow down the carbonation process towards the rebar. The minimum concrete cover will depend on the environmental conditions encountered and must be thicker when the concrete is also exposed to moisture and chloride (proximity to the sea, use of de-icing salt for bridges or roads, ...). A high quality concrete made with a low water-to-cement (w/c) ratio will have a lower porosity and will be less permeable to water and to the ingress of corrosive species (dissolved oxygen, chloride, ...). A thicker cover or a more compact concrete will also reduce the diffusion of CO_2 in the concrete, protecting it better from carbonation and maintaining a higher pH for a longer time period, increasing so the rebar service life.

Guidelines

National codes also specify minimum cover requirements based on their respective local exposure conditions.

Table : Other National Concrete Cover Requirements

Country	*Concrete Code*	*Range of Concrete Cover (mm)*
UK	BS:8110	25-50
EU	EN 1992 (EC2)	diametre +10 - 55
USA	ACI:318	40-50
Australia	AS:3600	15-30

Paradox

Large cover depths (50–75 mm) are required to protect reinforcement against corrosion in aggressive environments, but thick cover leads to increased crack widths in flexural reinforced concrete members. Large crack-widths (greater than 0.3 mm) permit ingress of moisture and chemical attack to the concrete, resulting in possible corrosion of reinforcement and deterioration of concrete. Therefore, thick covers defeat the very purpose for which it is provided. There is a need for judicious balance of cover depth and crack width requirements.

A possible economical solution for this paradox is the placing of a second layer of corrosion-resistant reinforcement like stainless steel rebars or meshes or FRP rebars in the concrete cover to distribute the cracks.

Formwork

Formwork is the term given to either temporary or permanent molds into which concrete or similar materials are poured. In the context of concrete construction, the falsework supports the shuttering moulds.

Figure : Modular steel frame formwork for a foundation

Figure : Timber formwork for a concrete column

Figure : Re-usable Plastic-Formwork for mass housing

Formwork and Concrete Form Types

Formwork comes in several types:

1. *Traditional timber formwork.* The formwork is built on site out of timber and plywood or moisture-resistant particleboard. It is easy to produce but time-consuming for larger structures, and the plywood facing has a relatively short lifespan. It is still used extensively where the labour costs are lower than the costs for procuring re-usable formwork. It is also the most flexible type of formwork, so even where other systems are in use, complicated sections may use it.
2. *Engineered Formwork System.* This formwork is built out of prefabricated modules with a metal frame (usually steel or aluminium) and covered on the application (concrete) side with material having the wanted surface structure (steel, aluminium, timber, etc.). The two major advantages of formwork systems, compared to traditional timber formwork, are speed of construction (modular systems pin, clip, or screw together quickly) and lower life-cycle costs (barring major force, the frame is almost indestructible, while the covering if made of wood; may have to be replaced after a few - or a few dozen - uses, but if the covering is made with steel or aluminium the form can achieve up to two thousand uses depending on care and the applications).
3. *Re-usable plastic formwork.* These interlocking and modular systems are used to build widely variable, but relatively simple, concrete structures. The panels are lightweight and very robust. They are especially suited for low-cost, mass housing schemes.
4. *Permanent Insulated Formwork.* This formwork is assembled on site, usually out of insulating concrete forms (ICF). The formwork stays in place after the concrete has cured, and may provide advantages in terms of speed, strength, superior thermal and acoustic insulation, space to run utilities within the EPS layer, and integrated furring strip for cladding finishes.
5. *Stay-In-Place structural formwork systems.* This formwork is assembled on site, usually out of prefabricated fibre-reinforced plastic forms. These are in the shape of hollow tubes, and are usually used for columns and piers. The formwork stays in place after the concrete has cured and acts as axial and shear reinforcement, as well as serving to confine the concrete and prevent against environmental effects, such as corrosion and freeze-thaw cycles.

Slab Formwork (Deck Formwork)

Figure : Pantheon dome

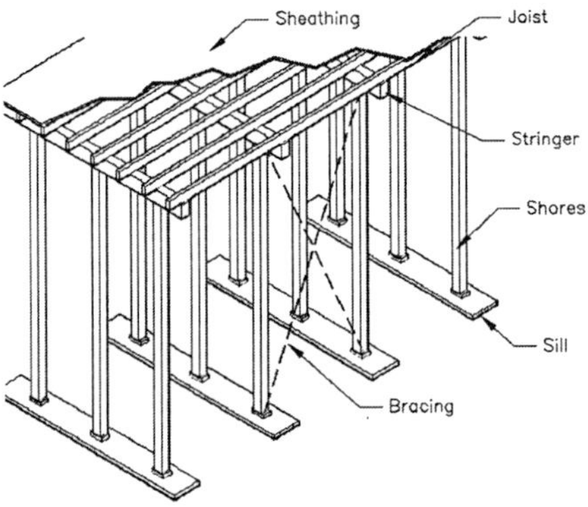

Figure : Schematic sketch of traditional formwork

Figure : Modular formwork with deck for housing project in Chile

History

Some of the earliest examples of concrete slabs were built by Roman engineers. Because concrete is quite strong in resisting compressive loads, but has relatively poor Tensile or torsional strength, these early structures consisted of arches, vaults and domes. The most notable concrete structure from this period is the Pantheon in Rome.

To mould these structure, temporary scaffolding and formwork or falsework was built in the future shape of the structure. These building techniques were not isolated to pouring concrete, but were and are widely used in Masonry.

Because of the complexity and the limited production capacity of the building material, concrete's rise as a favoured building material did not occur until the invention of Portland cement (and developments by the Edison Portland Cement Company) and reinforced concrete.

Timber Beam Slab Formwork

Similar to the traditional method, but stringers and joist are replaced with engineered wood beams and supports are replaced with metal props. This makes this method more systematic and reusable.

Traditional Slab Formwork

On the dawn of the rival of concrete in slab structures, building techniques for the temporary structures were derived again from masonry and carpentry. The traditional slab formwork technique consists of supports out of lumber or young tree trunks, that support rows of stringers assembled roughly 3 to 6 feet or 1 to 2 metres apart, depending on thickness of slab. Between these stringers, joists are positioned roughly 12 inches, 30 centimetres apart upon which boards or plywood are placed.

The stringers and joists are usually 4 by 4 inch or 4 by 6 inch lumber. The most common imperial plywood thickness is ¾ inch and the most common metric thickness is 21 mm.

Metal Beam Slab Formwork

Similar to the traditional method, but stringers and joist are replaced with aluminium forming systems or steel beams and supports are replaced with metal props. This also makes this method more systematic and reusable.

Modular Slab Formwork

These systems consist of prefabricated timber, steel or aluminium beams and formwork modules. Modules are often no larger than 3 to 6 feet or 1 to 2 metres in size. The beams and formwork are typically set by hand and pinned, clipped, or screwed together. The advantages of a modular system are: does not require a crane to place the formwork, speed of construction with unskilled labour, formwork modules can be removed after concrete sets leaving only beams in place prior to achieving design strength.

Table or Flying Form Systems

These systems consist of slab formwork "tables" that are reused on multiple stories of a building without being dismantled. The assembled sections are either lifted per elevator or "flown" by crane from one story to the next. Once in position the gaps between the tables or table and wall are filled with "fillers". They vary in shape and size as well as their building material. The use of these systems can greatly reduce the time and manual labour involved in setting and striking the formwork. Their advantages are best utilised by large area and simple structures. It is also common for architects and engineers to design building around one of these systems.

Structure

A table is built pretty much the same way as a beam formwork but the single parts of this system are connected together in a way that makes them transportable. The most common sheathing is plywood, but steel and fibreglass are also in use. The joists are either made from timber, wood I-beams, aluminium or steel. The Stringers are sometimes made of wood I-beams but usually from steel channels. These are fastened together (screwed, weld or bolted) to become a "duck". These decks are usually rectangular but can also be other shapes.

Support

All support systems have to be height adjustable to allow the formwork to be placed at the correct height and to be removed after the concrete is cured. Normally adjustable metal props similar to (or the same as) those used by beam slab formwork are used to support these systems. Some systems combine stringers and supports into steel or aluminium trusses. Yet other systems use metal frame shoring towers, which the decks are attached to. Another common method is to attach the formwork decks to previously cast walls or columns, thus eradicating the use of vertical props altogether. In this method, adjustable support shoes are bolted through holes (sometimes tie holes) or attached to cast anchors.

Size

The size of these tables can vary from 70 to 1,500 square feet (6.5 to 140 m^2). There are two general approaches in this system.

Crane Handled: this approach consists of assembling or producing the tables with a large formwork area that can only be moved up a level by crane. Typical widths can be 15, 18 or 20 ft. or 5 to 7 metres but their width can be limited, so that it is possible to transport them assembled, without having to pay for an oversize load. The length vary and can be up to 100 ft. (or more) depending on the crane capacity. After the concrete is cured, the decks are lowered and moved with rollers or trolleys to the edge of the building. From then on the protruding side of the table is lifted by crane whiles the rest of the table is rolled out of the building. After the centre of gravity is outside of the building the table reattached to another crane and flown to the next level or position.

This technique is fairly common in the United States and east Asian countries. The advantages of this approach are the further reduction of manual labour time and cost per area of slab and a simple and systematic building technique. The disadvantages of this approach are the necessary high lifting capacity of building site cranes, additional expensive crane time, higher material costs and little flexibility.

By this approach the tables are limited in size and weight. Typical widths are between 6 to 10 ft or 2 to 3 metres, typical lengths are between 12 and 20 ft or 4 to 7 metres, though table sizes may vary in size and form. The major distinction of this approach is that the tables are lifted either with a crane transport fork or by material platform elevators attached to the side of the building. They are usually transported horizontally to the elevator or crane lifting platform singlehandedly with shifting trolleys depending on their size and construction. Final positioning adjustments can be made by trolley. This technique enjoys popularity in the US, Europe and generally in high labour cost countries. The advantages of this approach in comparison to beam formwork or modular formwork is a further reduction of labour time and cost. Smaller tables are generally easier to customize around geometrically complicated buildings, (round or non rectangular) or to form around columns in comparison to their large counterparts. The disadvantages of this approach are the higher material costs and increased crane time (if lifted with crane fork).

Usage

For removable forms, once the concrete has been poured into formwork and has set (or *cured*), the formwork is *struck* or *stripped* (removed) to expose the finished concrete. The time between pouring and formwork stripping depends on the job specifications, the cure required, and whether the form is supporting any weight, but is usually at least 24 hours after the pour is completed. For example, the California Department of Transportation requires the forms to be in place for 1–7 days after pouring, while the Washington State Department of Transportation requires the forms to stay in place for 3 days with a damp blanket on the outside.

Spectacular accidents have occurred when the forms were either removed too soon or had been under-designed to carry the load imposed by the weight of the uncured concrete. Less critical and much more common (though no less embarrassing and often costly) are those cases in which undersigned formwork bends or breaks during the filling process (especially if filled with a high-pressure concrete pump). This then results in fresh concrete escaping out of the formwork in a *form blowout*, often in large quantities.

Concrete exerts less pressure against the forms as it hardens, so forms are usually designed to withstand a number of feet per hour of pour rate to give the concrete at the bottom time to firm up. For example, wall or column forms are commonly designed for a pour rate between 4–8 ft/hr. The hardening is an asymptotic process, meaning that most of the final strength will be achieved after a short time, though some further hardening can occur depending on the cement type and admixtures.

Wet concrete also applies hydrostatic pressure to formwork. The pressure at the bottom of the form is therefore greater than at the top. In the illustration of the column formwork to the right, the 'column clamps' are closer together at the bottom. Note that the column is braced with steel adjustable 'formwork props' and uses 20 mm 'through bolts' to further support the long side of the column.

Moladi

Moladi is a South African construction company specialising in a reusable plastic formwork for use in affordable housing and low cost housing projects, mainly in third world countries. The process involves creating a mould the form of the complete house. This wall mould is then filled with an aerated form of mortar. The process is also claimed to be faster than traditional methods of construction.

The process has won the *Design for Development* award of the South African Bureau of Standards Design Institute in 1997, with the institute praising Moladi as:

> *...an interlocking and modular shutter system for moulding complex concrete structures. The panels are lightweight and very robust. The system is especially suited for affordable low-cost, mass housing schemes.*

In addition to being a part of the drive by the South African government to replace shantytowns with proper houses, Moladi also exports to third world countries like Panama, where the company also plans to set up a factory. The company also opened a new factory in Port Elizabeth in 2008.

Climbing Formwork

Figure : Climbing formwork on a future residential skyscraper in Takapuna, New Zealand - the whole white upperstructure is actually formwork and associated working facilities.

Climbing formwork is a special type of formwork for vertical concrete structures that rises with the building process. While relatively complicated and costly, it can be an effective solution for buildings that are either very repetitive in form (such as towers or skyscrapers) or that require a seamless wall structure (using gliding formwork, a special type of climbing formwork).

Various types of climbing formwork exist, which are either relocated from time to time, or can even move on their own (usually on hydraulic jacks, required for self-climbing and gliding formworks).

Process

Best known in the construction of towers, skyscrapers and other tall vertical structures, it allows the reuse of the same formwork over and over and over for identical (or very similar) sections / stories further up the structure. It can also enable very large concrete structures to be constructed in one single pour (which may take days or weeks as the formwork rises with the process), thus creating seamless structures with enhanced strength and visual appearance, as well as reducing construction times and material costs (at the joints which would otherwise require extra reinforcement / connectors).

The climbing formwork structure normally does not only contain the formwork itself, but also usually provides working space / scaffolds for construction crews. It may also provide areas for machinery and screens for weather protection, up to being fully enclosed while yet staying modular around a changing building structure.

Types

- Climbing formwork (crane-climbing) - in this type of climbing formwork, the formwork around the structure is displaced upwards with the help of one or more cranes once the hardening of the concrete has proceeded far enough. This may entail lifting the whole section, or be achieved segmentally.
- Climbing formwork (self-climbing) - In this type of formwork, the structure elevates itself with the help of mechanic leverage equipment (usually hydraulic). To do this, it is usually fixed to sacrificial cones or rails emplaced in the previously cast concrete.
- Gliding formwork - This type of formwork is similar to the self-climbing type above. However, the climbing process is continuous instead of intermittent, and is usually only interrupted for a very short time (for example to fix the mounting mechanisms to new anchoring points). The advantage is that it will produce seamless structures, but it requires a continuous, uninterrupted process throughout, with serious potential quality and stability problems if the pour has to be stopped.

Skyscraper

A skyscraper is a tall, continuously habitable building of many stories, often designed for office and commercial use. There is no official definition or height above which a building may be classified as a skyscraper.

One common feature is that skyscrapers tend to make use of a steel framework structure from which walls are suspended, rather than having load-bearing walls as seen in conventional buildings.

As there is no official definition of what constitutes a skyscraper, a relatively small building may be considered one if it protrudes well above its built environment and changes the overall skyline. The maximum height of structures has progressed historically with building methods and technologies. 'Supertall' has arisen as a contemporary expression for exceptionally tall buildings, although again there is no formal definition.

Definition

The word "skyscraper" originally was a nautical term referring to a small triangular sail set above the skysail on a sailing ship. The term was first applied to buildings of steel framed construction of at least 10 storeys in the late 19th century, a result of public amazement at the tall buildings being built in major cities like Chicago, New York City, Detroit, and St. Louis. The first steel frame skyscraper was the Home Insurance Building (originally 10 storeys with a height of 42 m or 138 ft) in Chicago, Illinois in 1885. Some point to New York's seven-floor Equitable Life Assurance Building, built in 1870, as an early skyscraper for its innovative use of a kind of skeletal frame, but such designation depends largely on what factors are chosen. Even the scholars making the argument find it to be purely academic.

The structural definition of the word *skyscraper* was refined later by architectural historians, based on engineering developments of the 1880s that had enabled construction of tall multi-storey buildings. This definition was based on the steel skeleton—-as opposed to constructions of load-bearing masonry, which passed their practical limit in 1891 with Chicago's Monadnock Building. The steel frame developed in stages of increasing self-sufficiency, with several buildings in Chicago and New York advancing the technology that allowed the steel frame to carry a building on its own. Today, however, many of the tallest skyscrapers are built almost entirely with reinforced concrete.

Skyscraper and Supertall

The Emporis Standards Committee defines a high-rise building as "a multi-storey structure between 35–100 metres tall, or a building of unknown height from 12–39 floors" and a skyscraper as "a multi-storey building whose architectural height is at least 100 metres." Some structural engineers define a highrise as any vertical construction for which wind is a more significant load factor than earthquake or weight. Note that this criterion fits not only high-rises but some other tall structures, such as towers.

The word *skyscraper* often carries a connotation of pride and achievement. The skyscraper, in name and social function, is a modern expression of the age-old symbol of the world centre or *axis mundi*: a pillar that connects earth to heaven and the four compass directions to one another.

A loose convention of some in the United States and Europe draws the lower limit of a skyscraper at 150 metres (~500 ft).

History

Figure : The Two Towers of Bologna in the 12th century reached 97.2 m in height.

Before the 19th Century

Modern skyscrapers are built with materials such as steel, glass, reinforced concrete and granite, and routinely utilise mechanical equipment such as water pumps and elevators. Until the 19th century, buildings of over six stories were rare, as having great numbers of stairs to climb was impractical for inhabitants, and water pressure was usually insufficient to supply running water above 50 m (164 ft).

The tallest building in ancient times was the Great Pyramid of Giza in ancient Egypt, which was 146 metres (479 ft) tall and was built in the 26th century BC. Its height was not surpassed for thousands of years, possibly until the 14th century AD with the construction of Lincoln Cathedral (though its height is disputed), which in turn was not surpassed in height until the Washington Monument in 1884. However, being uninhabited buildings, none of these buildings actually complies with the definition of a skyscraper.

High-rise apartment buildings already flourished in classical antiquity: ancient Roman insulae in Rome and other imperial cities reached up to 10 and more stories. Several emperors, beginning with Augustus (r. 30 BC-14 AD), attempted to establish limits of 20–25 m for multi-storey buildings, but met with only limited success. The lower floors were typically occupied by either shops or wealthy families, while the upper stories were rented out to the lower classes. Surviving Oxyrhynchus Papyri indicate that seven-storey buildings even existed in provincial towns, such as in 3rd century AD Hermopolis in Roman Egypt.

The skylines of many important medieval cities had large numbers of high-rise urban towers. Wealthy families built these towers for defensive purposes and as status symbols. The residential Towers of Bologna in the 12th century, for example, numbered between 80 to 100 at a time, the largest of which (known as the "Two Towers") rise to 97.2 metres (319 ft). In Florence, a law of 1251 decreed that all urban buildings should be reduced to a height of less than 26 m, the regulation immediately put into effect. Even medium-sized towns at the time such as San Gimignano are known to have featured 72 towers up to 51 m height.

The medieval Egyptian city of Fustat housed many high-rise residential buildings, which Al-Muqaddasi in the 10th century described as resembling minarets. Nasir Khusraw in the early 11th century described some of them rising up to 14 stories, with roof gardens on the top floor complete with ox-drawn water wheels for irrigating them. Cairo in the 16th century had high-rise apartment buildings where the two lower floors were for commercial and storage purposes and the multiple stories above them were rented out to tenants. An early example of a city consisting entirely of high-rise housing is the 16th-century city of Shibam in Yemen. Shibam was made up of over 500 tower houses, each one rising 5 to 11 storeys high, with each floor being an apartment occupied by a single family. The city was built in this way in order to protect it from Bedouin attacks. Shibam still has the tallest mudbrick buildings in the world, with many of them over 30 m (98 ft) high.

An early modern example of high-rise housing was in 17th-century Edinburgh, Scotland, where a defensive city wall defined the boundaries of the city. Due to the restricted land area available for development, the houses increased in height instead. Buildings of 11 stories were common, and there are records of buildings as high as 14 stories. Many of the stone-built structures can still be seen today in the old town of Edinburgh.

The oldest iron framed building in the world, although only partially iron framed, is The Flaxmill (also locally known as the "Maltings"), in Shrewsbury, England. Built in 1797, it is seen as the "grandfather of skyscrapers", since its fireproof combination of cast iron columns and cast iron beams developed into the modern steel frame that made modern skyscrapers possible. Unfortunately, it lies derelict and needs much investment to keep it standing.

Early Skyscrapers

An early development was Oriel Chambers in Liverpool. Designed by local architect Peter Ellis in 1864, the building was the world's first iron-framed, glass curtain-walled office building. It was only 5 floors high. Further developments led to the world's first skyscraper, the ten-storey Home Insurance Building in Chicago, built in 1884–1885. While its height is not considered very impressive today, it was at that time. The architect, Major William Le Baron Jenney, created a load-bearing structural frame. In this building, a steel frame supported the entire weight of the walls, instead of load-bearing walls carrying the weight of the building. This development led to the "Chicago skeleton" form of construction.

Burnham and Root's 1889 Rand McNally Building in Chicago, 1889, was the first all-steel framed skyscraper, while Louis Sullivan's Wainwright Building in St. Louis, Missouri, 1891, was the first steel-framed building with soaring vertical bands to emphasize the height of the building and is therefore considered by some to be the first true skyscraper.

Most early skyscrapers emerged in the land-strapped areas of Chicago, London, and New York toward the end of the 19th century. A land boom in Melbourne, Australia between 1888–1891 spurred the creation of a significant number of early skyscrapers, though none of these were steel reinforced and few remain today. Height limits and fire restrictions were later introduced. London builders soon found building heights limited due to a complaint from Queen Victoria, rules that continued to exist with few exceptions until the 1950s.

Concerns about aesthetics and fire safety had likewise hampered the development of skyscrapers across continental Europe for the first half of the twentieth century (with the notable exceptions of the 1898 Witte Huis *(White House)* in Rotterdam; the Royal Liver Building in Liverpool, completed in 1911 and 90 m (300 ft) high; and the 17-storey Kungstornen *(Kings' Towers)* in Stockholm, Sweden, which were built 1924–25, the 15-storey Edificio Telefónica in Madrid, Spain, built in 1929; the 26-storey Boerentoren in Antwerp, Belgium, built in 1932; and the 31-storey Torre Piacentini in Genoa, Italy, built in 1940). After an early competition between Chicago and New York City for the world's tallest building, New York took the lead by 1895 with the completion of the American Surety Building, leaving New York with the title of tallest building for many years. New York City developers competed among themselves, with successively taller buildings claiming the title of "world's tallest" in the 1920s and early 1930s, culminating with the completion of the Chrysler Building in 1930 and the Empire State Building in 1931, the world's tallest building for forty years. The first completed World Trade Centre tower became the world's tallest building in 1972. However, it was soon overtaken by the Sears Tower (now Willis Tower) in Chicago within two years. The Sears Tower stood as the world's tallest building for 24 years, from 1974 until 1998, until it was edged out by Petronas Twin Towers in Kuala Lumpur, which held the title for six years.

Modern Skyscrapers

From the 1930s onwards, skyscrapers also began to appear in Latin America (São Paulo, Santiago, Caracas, Bogotá, Mexico City) and in Asia (Tokyo, Shanghai, Hong Kong, Manila, Singapore, Mumbai, Seoul, Kuala Lumpur, Taipei, Bangkok). Immediately after World War II, the Soviet Union planned eight massive skyscrapers dubbed "Stalin Towers" for Moscow; seven of these were eventually built.

The rest of Europe also slowly began to permit skyscrapers, starting with Madrid, during the 1950s. Finally, skyscrapers also began to be constructed in cities of Africa, the Middle East and Oceania (mainly Australia) from the late 1950s.

In the early 1960s structural engineer Fazlur Khan realised that the rigid steel frame structure that had "dominated tall building design and construction so long was not the only system fitting for tall buildings", marking "the beginning of a new era of skyscraper revolution in terms of multiple structural systems." His central innovation in skyscraper design and construction was the idea of the "tube" structural system, including the "framed tube", "trussed tube", and "bundled tube".

These systems allowed far greater economic efficiency, and also allowed efficient skyscrapers to take on various shapes, no longer needing to be box-shaped. Over the next fifteen years, many towers were built by Khan and the "Second Chicago School", including the massive 442-metre (1,451-foot) Willis Tower. Since 2000, Cities like Chicago , Shanghai , Dubai, New York, and Toronto have experienced a huge surge in skyscraper construction. Chicago, Hong Kong, and New York City, otherwise known as "the big three," are recognised in architectural circles as having especially compelling skylines.

A landmark skyscraper can inspire a boom of new high-rise projects in its city, as Taipei 101 has done in Taipei since its opening in 2004. In 2010, The Bank of America Tower at One Bryant Park became the world's first commercial LEED Platinum skyscraper.

History of Tallest Skyscrapers

At the beginning of the 20th century, New York City was a centre for the Beaux-Arts architectural movement, attracting the talents of such great architects as Stanford White and Carrere and Hastings. As better construction and engineering technology became available as the century progressed, New York and Chicago became the focal point of the competition for the tallest building in the world.

Each city's striking skyline has been composed of numerous and varied skyscrapers, many of which are icons of 20th century architecture:

- The Flatiron Building, designed by Daniel Hudson Burnham and standing 285 ft (87 m) high, was one of the tallest buildings in the city upon its completion in 1902, made possible by its steel skeleton. It was one of the first buildings designed with a steel framework, and to achieve this height with other construction methods of that time would have been very difficult. (The 1889 Tower Building, designed by Bradford Gilbert and considered by some to be New York's first skyscraper, may have been the first building to use a skeletal steel frame.) Subsequent buildings such as the Singer Building, the Metropolitan Life Tower were higher still.

- The Woolworth Building, a neo-Gothic "Cathedral of Commerce" overlooking City Hall, was designed by Cass Gilbert. At 792 feet (241 m), it became the world's tallest building upon its completion in 1913, an honour it retained until 1930, when it was overtaken by 40 Wall Street.
- That same year, the Chrysler Building took the lead as the tallest building in the world, scraping the sky at 1,046 feet (319 m). Designed by William Van Alen, an Art Deco style masterpiece with an exterior crafted of brick, the Chrysler Building continues to be a favourite of New Yorkers to this day.
- The Empire State Building, the first building to have more than 100 floors (it has 102), was completed the following year. It was designed by Shreve, Lamb and Harmon in the contemporary Art Deco style. The tower takes its name from the nickname of New York State. Upon its completion in 1931 at 1,250 feet (381 m), it took the top spot as tallest building, and towered above all other buildings until 1972. The antenna mast added in 1951 brought pinnacle height to 1,472 feet (449 m), lowered in 1984 to 1,454 feet (443 m).
- The World Trade Centre officially reached full height in 1972, was completed in 1973, and consisted of two tall towers and several smaller buildings. For a short time, the first of the two towers was the world's tallest building. Upon completion, the towers stood for 28 years, until the September 11 attacks destroyed the buildings in 2001. Various governmental entities, financial firms, and law firms called the towers home.
- The Willis Tower (formerly Sears Tower) was completed in 1974, one year after the World Trade Centre, and surpassed it as the world's tallest building. It was the first building to employ the "bundled tube" structural system, designed by Fazlur Khan. The building was not surpassed in height until the Petronas Towers were constructed in 1998, but remained the tallest in some categories until Burj Khalifa surpassed it in all categories in 2010. It is currently the tallest building in the United States.

Momentum in setting records passed from the United States to other nations with the opening of the Petronas Twin Towers in Kuala Lumpur, Malaysia, in 1998. The record for world's tallest building remained in Asia with the opening of Taipei 101 in Taipei, Taiwan, in 2004. A number of architectural records, including those of the world's tallest building and tallest free-standing structure, moved to the Middle East with the opening of the Burj Khalifa in Dubai, United Arab Emirates. This geographical transition is accompanied by a change in approach to skyscraper design. For much of the twentieth century large buildings took the form of simple geometrical shapes. This reflected the "international style" or modernist philosophy shaped by Bauhaus architects early in the century. The last of these, the Willis Tower and World Trade Centre towers in New York, erected in the 1970s, reflect the philosophy. Tastes shifted in the decade which followed, and new skyscrapers began to exhibit postmodernist influences. This approach to design avails itself of historical elements, often adapted and re-interpreted, in creating technologically modern structures. The Petronas Twin Towers recall Asian pagoda architecture and Islamic geometric principles. Taipei 101 likewise reflects the pagoda tradition as it incorporates ancient motifs such as the ruyi symbol.

Figure : Taipei 101, formerly the world's tallest skyscraper, was the first to exceed the half-kilometre mark.

Figure : The iconic World Trade Centre twin towers were destroyed in 2001.

Figure : The Willis Tower in Chicago was the world's tallest building from 1974 to 1998, and remains the tallest in the Western Hemisphere.

Figure : The Petronas Twin Towers.

Figure : The City of Capitals in Moscow is the tallest completed skyscraper in Europe.

Today

Today, skyscrapers are an increasingly common sight where land is expensive, as in the centres of big cities, because they provide such a high ratio of rentable floor space per unit area of land. They are built not just for economy of space; like temples and palaces of the past, skyscrapers are considered symbols of a city's economic power. Not only do they define the skyline, they help to define the city's identity.

Supertall Towers

At the time Taipei 101 broke the half-kilometre mark in height, it was already technically possible to build structures towering over a kilometre above the ground. Proposals for such structures have been put forward, including the Mile-High Tower to be built in Jeddah, Saudi Arabia and Burj Mubarak Al Kabir in Kuwait. Kilometre-plus structures present architectural challenges that may eventually place them in a new architectural category.

Future Notable Skyscrapers

The following skyscrapers, all contenders for being among the tallest in their city or region, are under construction and due to be completed in the next few years:

- Construction of the 133-floor, 640 m tall Digital Media City Landmark Building in Digital Media City, Seoul, South Korea, started in 2009, which will be the second-tallest building in the world when it is completed in 2015, housing the world's tallest observatory and hotels. Being constructed at the fastest speed among major skyscraper projects by South Korea's Samsung C&T (who also built Burj Khalifa), the supertall is the first skyscraper to contain an entire city inside a building, including the world's largest aquarium, a luxury department store, shopping malls, clinic centre, high-tech offices, first-class apartments, six to eight-star hotels, a concert restaurant, a broadcasting studio and an art centre.
- Construction of the Shanghai Tower started on 29 November 2008. The tower will be 632 m (2,073 ft) high and have 127 floors. The building will feature a glass curtain wall and nine indoor gardens when it is completed in 2014.
- Construction of the 151-floor, 610 m tall 151 Incheon Tower in Songdo International City, Incheon, South Korea, started in 2008, which will be the tallest twin towers in the world when it is completed in 2014.

- The Abraj Al-Bait Towers, also known as the "Mecca Royal Clock Hotel Tower" is a complex under construction in Mecca, Saudi Arabia by the Saudi Binladin Group. The complex consists of seven towers, and the tallest tower (Hotel Tower) will have a height of 601 m (1,972 ft). Upon completion in 2011, the structure will have the largest floor area of any structure in the world, at 1,500,000 square metres (16,137,600 sq ft).
- Construction of the 110-floor, 510 m tall Busan Lotte World, Busan, South Korea, started in 2009. It is due for completion in 2016.
- One World Trade Centre is currently under construction in New York City and will be the tallest tower in the redevelopment of the site of the former World Trade Centre. Its pinnacle will reach a height of 541.4 m (1,776 ft), a height (in feet) representing the year of the United States Declaration of Independence.
- The 308 m (1,010 ft) Tour Generali in Paris La Défense, scheduled to be completed in 2013, is an entirely green building office skyscraper that is set to be the tallest building in Paris and the second tallest in the European Union after The Shard in London.
- Construction of The Shard in London started in March 2009, and is scheduled to be completed in May 2012, in time for the London Olympics. At 310 m (1,017 ft), it is set to be the tallest building in the European Union.

Sustainability

The skyscraper as a concept is a product of the industrialized age, made possible by cheap energy and raw materials. The amount of steel, concrete and glass needed to construct a skyscraper is vast, and these materials represent a great deal of embodied energy. Tall skyscrapers are very heavy, which means that they must be built on a sturdier foundation than would be required for shorter, lighter buildings.

Building materials must also be lifted to the top of a skyscraper during construction, requiring more energy than would be necessary at lower heights. Furthermore, a skyscraper consumes a lot of electricity because potable and non-potable water must be pumped to the highest occupied floors, skyscrapers are usually designed to be mechanically ventilated, elevators are generally used instead of stairs, and natural lighting cannot be utilised in rooms far from the windows and the windowless spaces such as elevators, bathrooms and stairwells.

In the lower levels of a skyscraper a larger percentage of the building cross section must be devoted to the building structure and services than is required for lower buildings:-

- More structure – because it must be stronger to support more floors above
- The elevator conundrum creates the need for more lift shafts – everyone comes in at the bottom and they all have to pass through the lower part of the building to get to the upper levels.
- Building services – power and water enters the building from below and have to pass through the lower levels to get to upper levels.

In low-rise structures, the support rooms (chillers, transformers, boilers, pumps and air handling units) can be put in basements or roof space – areas which have low rental value. There is, however, a limit to how far this plant can be located from the area it serves. The farther away it is the larger the risers for ducts and pipes from this plant to the floors they serve and the more floor area these risers take. In practice this means that in highrise buildings this plant is located on 'plant levels' at intervals up the building.

Quotations

"What is the chief characteristic of the tall office building? It is lofty. It must be tall. The force and power of altitude must be in it, the glory and pride of exaltation must be in it. It must be every inch a proud and soaring thing, rising in sheer exaltation that from bottom to top it is a unit without a single dissenting line."

—Louis Sullivan's *The Tall Office Building Artistically Considered* (1896)

Chapter 2

Skyscraper Design and Construction

The design and construction of skyscrapers involves creating safe, habitable spaces in very tall buildings. The buildings must support their weight, resist wind and earthquakes, and protect occupants from fire. Yet they must also be conveniently accessible, even on the upper floors, and provide utilities and a comfortable climate for the occupants. The problems posed in skyscraper design are considered among the most complex encountered given the balances required between economics, engineering, and construction management.

Basic Design Considerations

Good structural design is important in most building design, but particularly for skyscrapers since even a small chance of catastrophic failure is unacceptable given the high price. This presents a paradox to civil engineers: the only way to assure a lack of failure is to test for all modes of failure, in both the laboratory and the real world. But the only way to know of all modes of failure is to learn from previous failures. Thus, no engineer can be absolutely sure that a given structure will resist all loadings that could cause failure, but can only have large enough margins of safety such that a failure is acceptably unlikely. When buildings do fail, engineers question whether the failure was due to some lack of foresight or due to some unknowable factor.

Loading and Vibration

The load a skyscraper experiences is largely from the force of the building material itself. In most building designs, the weight of the structure is much larger than the weight of the material that it will support beyond its own weight. In technical terms, the dead load, the load of the structure, is larger than the live load, the weight of things in the structure (people, furniture, vehicles, etc.). As such, the amount of structural material required within the lower levels of a skyscraper will be much larger than the material required within higher levels.

This is not always visually apparent. The Empire State Building's setbacks are actually a result of the building code at the time, and were not structurally required. On the other hand John Hancock Centre's shape is uniquely the result of how it supports loads. Vertical supports can come in several types, among which the most common for skyscrapers can be categorized as steel frames, concrete cores, tube within tube design, and shear walls.

The wind loading on a skyscraper is also considerable. In fact, the lateral wind load imposed on super-tall structures is generally the governing factor in the structural design. Wind pressure increases with height, so for very tall buildings, the loads associated with wind are larger than dead or live loads.

Other vertical and horizontal loading factors come from varied, unpredictable sources, such as earthquakes.

Shear Walls

A shear wall, in its simplest definition, is a wall where the entire material of the wall is employed in the resistance of both horizontal and vertical loads. A typical example is a brick or cinderblock wall. Since the wall material is used to hold the weight, as the wall expands in size, it must hold considerably more weight. Due to the features of a shear wall, it is acceptable for small constructions, such as suburban housing or an urban brownstone, to require low material costs and little maintenance. In this way, shear walls, typically in the form of plywood and framing, brick, or cinderblock, are used for these structures. For skyscrapers, though, as the size of the structure increases, so does the size of the supporting wall. Large structures such as castles and cathedrals inherently addressed these issues due to a large wall being advantageous (castles), or ingeniously designed around (cathedrals). Since skyscrapers seek to maximise the floor-space by consolidating structural support, shear walls tend to be used only in conjunction with other support systems.

Steel Frame

The classic concept of a skyscraper is a large steel box with many small boxes inside it. The genius of the steel frame is its simplicity. By eliminating the inefficient part of a shear wall, the central portion, and consolidating support members in a much stronger material, steel, a skyscraper could be built with both horizontal and vertical supports throughout. This method, though simple, has drawbacks. Chief among these is that as more material must be supported (as height increases), the distance between supporting members must decrease, which actually in turn, increases the amount of material that must be supported.

Tube Frame

Since 1963, a new structural system of framed tubes appeared. Fazlur Khan and J. Rankine defined the framed tube structure as "a three dimensional space structure composed of three, four, or possibly more frames, braced frames, or shear walls, joined at or near their edges to form a vertical tube-like structural system capable of resisting lateral forces in any direction by cantilevering from the foundation." Closely spaced interconnected exterior columns form the tube. Horizontal loads (primarily wind) are supported by the structure as a whole. About half the exterior surface is available for windows. Framed tubes allow fewer interior columns, and so create more usable floor space. Where larger openings like garage doors are required, the tube frame must be interrupted, with transfer girders used to maintain structural integrity. Tube-frame construction was first used in the DeWitt-Chestnut Apartment Building, designed by Khan and completed in Chicago in 1963. It was used soon after for the John Hancock Centre and in the construction of the World Trade Centre.

A variation on the tube frame is the bundled tube, which uses several interconnected tube frames. The Willis Tower in Chicago used this design, employing nine tubes of varying height to achieve its distinct appearance. The bundle tube design was not only highly efficient in economic terms, but it was also "innovative in its potential for versatile formulation of architectural space. Efficient towers no longer had to be box-like; the tube-units could take on various shapes and could be bundled together in different sorts of groupings." The bundled tube structure meant that "buildings no longer need be boxlike in appearance: they could become sculpture."

Tube structures have since been used in many other later skyscrapers, including the construction of the World Trade Centre, Petronas Towers, Jin Mao Building, and most other supertall skyscrapers since the 1960s. The strong influence of tube structure design is also evident in the construction of the current tallest skyscraper, the Burj Khalifa.

The Elevator Conundrum

The invention of the elevator was a precondition for the invention of skyscrapers, given that most people would not (or could not) climb more than a few flights of stairs at a time. The elevators in a skyscraper are not simply a necessary utility, like running water and electricity, but are in fact closely related to the design of the whole structure: a taller building requires more elevators to service the additional floors, but the elevator shafts consume valuable floor space. If the service core, which contains the elevator shafts, becomes too big, it can reduce the profitability of the building. Architects must therefore balance the value gained by adding height against the value lost to the expanding service core. Many tall buildings use elevators in a non-standard configuration to reduce their footprint. Buildings such as the former World Trade Centre Towers use Sky Lobbies, where express elevators take passengers to upper floors which serve as the base for local elevators. This allows architects and engineers to place elevator shafts on top of each other, saving space. Sky lobbies and express elevators take up a significant amount of space, however, and add to the amount of time spent commuting between floors. Other buildings, such as the Petronas Towers, use double-deck elevators, allowing more people to fit in a single elevator, and reaching two floors at every stop. It is possible to use even more than two levels on an elevator, although this has never been done. The main problem with double-deck elevators is that they cause everyone in the elevator to stop when only people on one level need to get off at a given floor.

World's Littlest Skyscraper

The Newby-McMahon Building, commonly referred to as the world's littlest skyscraper, is located at 701 LaSalle Street (on the corner of Seventh and LaSalle Streets) in downtown Wichita Falls, Texas. This late Neoclassical style red brick and cast stone structure is 40 ft (12 m) tall, and its exterior dimensions are 18 ft (5.5 m) deep and 10 ft (3.0 m) wide. Its interior dimensions are approximately 12 ft (3.7 m) by 9 ft (2.7 m), or approximately 108 sq ft (10.0 m^2). Steep, narrow, internal stairways leading to the upper floors occupy roughly 25 percent of the interior area.

Reportedly the result of a fraudulent investment scheme by a confidence man, the Newby-McMahon Building was a source of great embarrassment to the city and its residents after its completion in 1919. During the 1920s, the Newby-McMahon Building was featured in Robert Ripley's *Ripley's Believe It or Not!* syndicated column as "the world's littlest skyscraper", a sobriquet that has stuck with it ever since. The Newby-McMahon Building is now part of the Depot Square Historic District of Wichita Falls, which has been declared a Texas Historic Landmark.

Background

A large petroleum reservoir was discovered just west of the city of Burkburnett, a small town in Wichita County, Texas in 1912. Burkburnett and its surrounding communities became boomtowns, experiencing explosive growth of their populations and economies. By 1918, approximately 20,000 new settlers had taken up residence around the lucrative oil field, and many Wichita County residents became wealthy virtually overnight. As people streamed into the local communities in search of high-paying jobs, the nearby city of Wichita Falls began to grow in importance. Though it initially lacked the necessary infrastructure for this sudden increase in economic and industrial activity, Wichita Falls was a natural choice to serve as the local logistical hub, being the seat of Wichita County. Because office space was lacking, major stock transactions and mineral rights deals were conducted on street corners and in tents that served as makeshift headquarters for the new oil companies.

Building Proposal and Blueprints

The Newby-McMahon Building is a one-story brick building located near the railroad depot in downtown Wichita Falls, built in 1906 by Augustus Newby (1855–1909), a director of the Wichita Falls and Oklahoma City Railway Company. The oil-rig construction firm of J.D. McMahon, a petroleum landman and structural engineer from Philadelphia, was one of seven tenants whose offices were based in the original Newby Building.

According to local legend, when McMahon announced in 1919 that he would build a highrise annex to the Newby Building as a solution to the newly wealthy city's urgent need for office space, investors were eager to seize the opportunity to become even wealthier. McMahon collected $200,000 in investment capital from this group of naïve investors, promising to construct a highrise office building across the street from the St. James Hotel. The proposed skyscraper depicted in the blueprints that he distributed was clearly labelled as being 480" tall and consisting of four floors. McMahon is said to have neglected to mention that the scale of his blueprints was in inches rather than feet.

Construction of Edifice and Ensuing Legal Battle

McMahon used his own construction crews to build the *McMahon Building* on the small, unused piece of property next to the Newby Building, without obtaining prior consent from the owner of the property, who lived in Oklahoma. As the building began to take shape, the investors realised they had been swindled into purchasing a four-story edifice that was only 40 ft (480 in) tall, rather than the 480 ft (150 m) structure they were expecting. At that time, the 792 ft (241 m) Woolworth Building in New York City was the tallest building in the world. They brought a lawsuit against McMahon, but to their dismay, the real estate and construction deal was declared legally binding by a local judge. As McMahon had built exactly according to the blueprints they'd signed off on, there was to be no legal remedy for the deceived investors. They did recover a small portion of their investment from the elevator company, which refused to honor the contract after they learned of the confidence trick. There was no stairway installed in the building upon its initial completion, as none was included in the original blueprints. Rather, a ladder was employed to gain access to the upper three floors. By the time construction was complete, McMahon had left Wichita Falls and perhaps even Texas, taking with him the balance of the investors' money.

Early Occupancy and Subsequent Abandonment

Upon its completion and opening in 1919, the Newby-McMahon Building was an immediate source of great embarrassment to the city and its residents. The ground floor had six desks representing the six different companies that occupied the building as its original tenants. Throughout most of the 1920s, the building housed only two firms. During the 1920s, the Newby-McMahon Building was featured in Robert Ripley's *Ripley's Believe It or Not!* syndicated column as "the world's littlest skyscraper," which is a name that has stuck with it ever since. The oil industry would ultimately prove to be a resource curse to Wichita Falls, and the Texas Oil Boom ended only a few years later. The building was vacated, boarded up, and virtually forgotten in 1929 as the Great Depression struck North Texas and office space became relatively inexpensive to lease or purchase. A fire gutted the building in 1931, rendering it unusable for a number of years.

After the Great Depression, the building housed a succession of tenants, including barber shops and cafés. The building changed hands many times and was scheduled for demolition on several occasions, but escaped this fate apparently because a sufficient number of local residents came to its defence. It was eventually deeded to the city of Wichita Falls. As the building continued to deteriorate, in 1986 the city gave the building to the Wichita County Heritage Society (WCHS), with the hope that it would eventually be restored, making it a viable part of the Depot Square Historic District.

Purchase and Renovation

By 1999, the Newby-McMahon Building had proved to be an excessive burden on the limited capital reserves of the WCHS. The following year, the city council hired the local architectural firm of Bundy, Young, Sims & Potter to stabilize the crumbling structure, amid steadily growing talk of demolishing the building. Dick Bundy and his partners became fascinated with the history and legacy of the building; they arranged a partnership with Marvin Groves Electric, another local business, to purchase the building. In December of 2000, the city council voted to allow the WCHS to sell the building to Marvin Groves for $3,748. On June 11, 2003, a storm swept through Wichita Falls, bringing gusts of wind as strong as 97 mph. A 15-foot section of brick wall from the McMahon Building complex was knocked down. The damage from this storm was repaired, but full restoration of the building and the adjacent Newby Building was delayed until late 2005. In June of that year, the City Council granted $25,000 in funds from the city's Tax Increment Financing Fund, to be invested in the restoration of the McMahon Building. Restoration of the building is estimated to have cost more than $254,000, the remainder of which was paid by the owners (Bundy, Young, Sims & Potter, Inc. and Marvin Groves Electric).

Current Status

With the passage of time, the Newby-McMahon Building has become a monument to a long-gone era. It has survived tornadoes, a fire, and decades of neglect to stand as a monument to the greed, graft, and gullibility of the oil boom days of North Texas. The building is currently part of the Depot Square Historic District of Wichita Falls, which has been declared a Texas Historic Landmark. The building has never met the criteria for the definition of a skyscraper, nor even that of a "highrise" building. Aside from serving as a local tourist attraction, the building is home to an antiques dealership, *The Antique Wood,* which opened in 2006 on the ground floor. The third floor has been converted into an artist's studio. The Newby-McMahon Building is among several historic buildings featured in the documentary film *Wichita Falls: The Future of Our Past,* a retrospective analysis of the city's architectural past produced in 2006 by Barry Levy, a public information officer with the city of Wichita Falls.

Skyscraper Index

The Skyscraper Index is a concept put forward in January 1999 by Andrew Lawrence, research director at Dresdner Kleinwort

Wasserstein, which showed that the world's tallest buildings have risen on the eve of economic downturns. Business cycles and skyscraper construction correlate in such a way that investment in skyscrapers peaks when cyclical growth is exhausted and the economy is ready for recession.

The buildings may actually be completed after the onset of the recession or later, when another business cycle pulls the economy up, or even cancelled. Unlike earlier instances of similar reasoning ("height is a barometre of boom"), Lawrence used skyscraper projects as a predictor of economic crisis, not boom.

Details

Lawrence started his paper, *The Skyscraper Index: Faulty Towers*, as a joke (emphasized by a title referencing a comedy show) and based his "index" on mere comparison of historical data, primarily from the United States experience. He dismissed overall construction and investment statistics, focusing only on record-breaking projects. The first notable example was the Panic of 1907. Two record-breaking skyscrapers, the Singer Building and Metropolitan Life Insurance Company Tower, were launched in New York before the panic and completed in 1908 and 1909, respectively. Met Life remained the world's tallest building until 1913. Another string of supertall towers — 40 Wall Street, Chrysler Building, Empire State Building — was launched shortly before to the Wall Street Crash of 1929.

The next record holders, World Trade Centre towers and Sears Tower, opened up in 1973, during the 1973–1974 stock market crash and the 1973 oil crisis. The last example available to Lawrence, Petronas Twin Towers, opened up in the wake of the 1997 Asian Financial Crisis and held the world height record for five years. Lawrence linked the phenomenon to overinvestment, speculation and monetary expansion but did not elaborate these underlying issues. The concept was revived in 2005, when Fortune warily observed five media corporations investing in new skyscrapers on Manhattan (none of them, including the tallest New York Times Building, broke any records).

The intuitively simple concept, publicized by business press in 1999, has been cross-checked within the framework of the Austrian Business Cycle Theory, itself borrowing on Richard Cantillon's eighteenth-century theories. Thornton (2005) listed three *Cantillon effects* that make skyscraper index valid. First, a decline in interest rates at the onset of a boom drives land prices. Second, a decline in interest rates allows increase in average size of a firm, creating demand for larger office spaces. Third, low interest rates provide investment to construction technologies that enable developers to break earlier records. All three factors peak at the end of growth period. Critics dismissed the skyscraper index as an unreliable tool: the post-World War I recession, recession of 1937 and the early 1980s recession were not marked by any record-breaking projects. Construction of Woolworth Building (world height record 1913–1930) was marked by a local overbuilding crisis in New York City in 1913–1915 concurrent with a record construction boom in Chicago. Thornton argues that completion of Woolworth Building was followed by a third-worst-ever quarterly decline in gross domestic product, thus it should not be considered an exception from the rule (as Lawrence himself did).

Cyclical patterns in real estate have been thoroughly studied before Lawrence, notably by Homer Hoyt in 1930s. A 1995 analysis of New York and Chicago experience by Carol Willis estimated that historically, two-thirds to three-quarters of skyscrapers were conceived for rent alone; corporate "edifices" imposing their owners brand name (including most of historical record-holders) were a minority, and they too leased space to tenants. Speculative real estate markets cycle between two different behaviour patterns. In "normal" times when value of resources is predictable, performance of a building project can be estimated reliably through well-tested formulae. In boom times, rational pricing gives way to irrational buyers' behaviour; buyers bet on ever-increasing demand and rents and are willing to pay more than they would normally. Willis said that "height is a barometre of boom", "the tallest building generally appear before the end of a boom, their height driven up by the speculative fever that affects both developers and lenders", citing cyclically inflated land values as the principal factor for increases in building height, but did not elevate this fact to become an "index".

A related concept, Skyscraper Indicator, was popularized by Ralph Nelson Elliott in 1930s. In some ways this appears to be an elaboration of C Northcote Parkinson's theory that only organisations in decline have sleek, well-planned buildings. His favourite example was, not a skyscraper, but the city of New Delhi (particularly the area now referred to as Lutyen's Delhi) - built shortly before India became independent of the British builders. The recently constructed Burj Khalifa may join this list. In October 2009, construction company Emaar announced that it had completed the exterior of the building;

within two months, the Dubai government came close to defaulting on its loans. Stephen Bayley from *The Daily Telegraph* commented, "For all the ambition of its construction, Dubai's new Khalifa Tower is a frightening, purposeless monument to the subprime era."

History of the Tallest Buildings in the World

Through history, the title of "world's tallest building" has been borne by various buildings. Since the invention of the skyscraper, New York City has been home to the world's tallest building for roughly 87 years (inside the period of 1870-1974, over nearly a dozen successive buildings). Chicago accumulated 30 years. This distinction was held exclusively within the United States for nearly 130 years (1870-1998) before returning to the eastern hemisphere. Preceding the current era of commercial skyscrapers, there was an era where the tallest buildings were Christian churches/cathedrals (1200s-1870), dominated by England and Germanic territories. And the era preceding this was the thousands of years where the Great Pyramid in Cairo was the tallest structure in the world. Other claims are made to "tallest structure" during the latter two eras, depending upon definitions applied, notably the Washington Monument (1884), which was surpassed in height by the Eiffel Tower in Paris (1889).

Definition of Terms

Meaning of "Building"

The earliest structures now known to be the tallest in the world were the Egyptian pyramids, with the Great Pyramid of Giza, at an original height of 146.5 metres (481 ft), being the tallest man–made. From then until the completion of the Washington Monument (capped in 1884) the world's tallest buildings were churches or cathedrals. Later, the Eiffel Tower and, still later, some radio masts and television towers were the world's tallest structures.

However, though all of these are *structures*, some are not *buildings* in the sense of being regularly inhabited or occupied. It is in this sense of being regularly inhabited or occupied that the term "building" is generally understood to mean when determining what is the world's tallest building. The non-profit international organisation Council on Tall Buildings and Urban Habitat (CTBUH), which maintains a set of criteria for determining the height of tall buildings, defines "building" as "[A] structure that is designed for residential, business or manufacturing purposes" that "has floors". Tall churches and cathedrals occupy a middle ground: their lower areas are regularly occupied, but much of their height is in bell towers and spires which aren't. Whether a church or cathedral is a "building" or merely a "structure" for the purposes of determining the title of "world's tallest building" is a subjective matter of definition.

Determination of Height

The Council on Tall Buildings and Urban Habitat uses three different criteria for determining the height of a tall building, each of which may give a different result. "Height of the highest floor" is one criterion, and "height to the top of any part of the building" is another, but the default criterion used by the CTBUH is "height of the architectural top of the building", which includes spires but not antennae, masts, or flag poles.

Tallest Buildings (1200s–1901)

Churches and Cathedrals

From 1200s until 1901, the world's tallest building was always a church or cathedral. In 1200s Old St Paul's Cathedral with its spire was completed. Completed in early 1300s, the central spire of Lincoln Cathedral surpassed it.

In 1549 this spire collapsed, thus making the shorter St. Olaf's church in Tallinn the tallest building. In 1625, the spire of this church burnt down, thus making the shorter St. Mary's Church in Stralsund the world's tallest building. In 1647, the bell tower of this church burned down, thus making the slightly shorter Strasbourg Cathedral the world's tallest building.

Years tallest	Name	Location	Height	Increase
1200s–1300*	Old St Paul's Cathedral	London	149 metres (489 ft)	1.66%
1300–1549*	Lincoln Cathedral	Lincoln	159.7 metres (524 ft)	7.16%
1549–1625	St. Olaf's Church	Tallinn	159 metres (522 ft)	-0.38%
1625–1647	St. Mary's Church	Stralsund	151 metres (495 ft)	-5.53%
1647–1874	Strasbourg Cathedral	Strasbourg	142 metres (466 ft)	-11.07%
1874–1876	Church of St. Nicholas	Hamburg	147 metres (482 ft)	-8.02%
1876–1880	Rouen Cathedral	Rouen	151 metres (495 ft)	-5.53%
1880–1890	Cologne Cathedral	Cologne	157.38 metres (516.3 ft)	-1.47%
1890–1901	Ulm Minster	Ulm	161.53 metres (530.0 ft)	1.15%

* - Also set record at time of completion as tallest *structure* ever built.

The 524 feet (160 m) height of Lincoln Cathedral is disputed by some, but accepted by most sources. The completion date for the spire is given as 1311 rather than 1300 by some sources. Also the 489 feet (149 m) height of the spire of Old St Paul's Cathedral, destroyed by lightning in 1561, is disputed, for example Christopher Wren (1632–1723) judged that an overestimate and gave height 460 feet (140 m).

Skyscrapers

In the 19th Century, a new kind of structure was developed, using an iron or steel internal structure (instead of the outer walls) to bear the building's weight. The taller of these buildings are called skyscrapers.

There is no one building that can be definitely termed the first skyscraper. The Equitable Life Building in New York was completed in 1870. At 7 storeys and 40 metres, it was the first office building to feature passenger elevators and was, at the time of its completion, the tallest non-church building in the world. Some consider the Equitable Life Building with its world-first elevators to be the first skyscraper, while others point to Chicago's Home Insurance Building with its innovation in using a steel-frame construction. Both design features would become standard in skyscrapers.

The buildings that were the tallest skyscrapers — but still shorter than the tallest church or cathedral — were:

Years tallest	*Name*	*Location*	*Height*	*Increase*
1870–1884	Equitable Life Building	New York	40 metres (130 ft)	-
1884–1890	Home Insurance Building	Chicago	42 metres (138 ft)	6.15%
1890–1894	New York World Building	New York	94 metres (308 ft)	136.92%
1894–1895	Manhattan Life Insurance Building	New York	100 metres (330 ft)	7.14%
1895–1899	Milwaukee City Hall	Milwaukee	108 metres (354 ft)	7.27%
1899–1901	Park Row Building	New York	119 metres (390 ft)	10.17%

Tallest Buildings (from 1901)

At 167 metres, the 1901 Philadelphia City Hall was taller than Ulm Minster and thus the tallest building of any kind. Since 1901, the world's tallest building has always been a secular skyscraper. The tallest *structure* at the beginning of this period was the Eiffel Tower, which was not surpassed until 1930 with the Chrysler Building.

Years tallest	*Name*	*Location*	*Height*	*Increase*
1901–1908	Philadelphia City Hall	Philadelphia	167 metres (548 ft)	3.40%
1908–1909	Singer Building	New York	186.57 metres (612.1 ft)	11.70%
1909–1913	Metropolitan Life Tower	New York	213.36 metres (700.0 ft)	14.36%
1913–1930	Woolworth Building	New York	241 metres (791 ft)	13.00%
1930	Bank of Manhattan Trust Building	New York	283 metres (928 ft)	17.32%
1930–1931*	Chrysler Building	New York	319.9 metres (1,050 ft)	13.15%
1931–1972*	Empire State Building	New York	381 metres (1,250 ft)	19.05%
1972–1974	World Trade Center #1	New York	417 metres (1,368 ft)	9.44%
1974–1998	Sears Tower	Chicago	442 metres (1,450 ft)	5.99%
1998–2003	Petronas Towers	Kuala Lumpur	451.9 metres (1,483 ft)	2.28%
2003–2010	Taipei 101	Taipei	509.2 metres (1,671 ft)	12.68%
2010–present*	Burj Khalifa	Dubai	828 metres (2,717 ft)	62.60%

* - Also set record at time of completion as tallest *structure* ever built.

The list of tallest buildings is based on the Council on Tall Buildings and Urban Habitat (CTBUH) default criteria of measuring to the highest architectural element. Other criteria would generate a different list. Shanghai World Financial Centre is not on the above list, but it surpassed Tapei 101 in 2008 to become the building with the highest occupied floor. Using the criteria of highest tip (including antennae), The World Trade Centre was the world's tallest building from 1972 to 2000, when the Sears Tower antenna was extended to give that building the world's tallest tip, a title it held until the 2010 completion of Burj Khalifa (Petronas Towers and Taipei 101 were never the world's tallest buildings by the highest–tip criteria).

Since 2010, Burj Khalifa in 2010 has been unquestionably the tallest building by any criteria. It has the highest architectural element, highest tip, and highest occupied floor, and is indeed the tallest structure of any kind in existence or ever built, surpassing the (now destroyed) 646.38 metres (2,120.7 ft) Warsaw Radio Mast. However, the 2008 Shanghai World Financial Centre still has the world's highest observation deck.

Since the completion of the Washington Monument in 1884, the world's tallest building has not usually also been the world's tallest structure. The exceptions are 1930–1954, when the Chrysler Building and then the Empire State building surpassed the Eiffel Tower (to be surpassed in turn in by a succession of broadcast masts, starting with the Griffin Television Tower in Oklahoma), and again from 2010 with the completion of Burj Khalifa.

Charts

1870-present

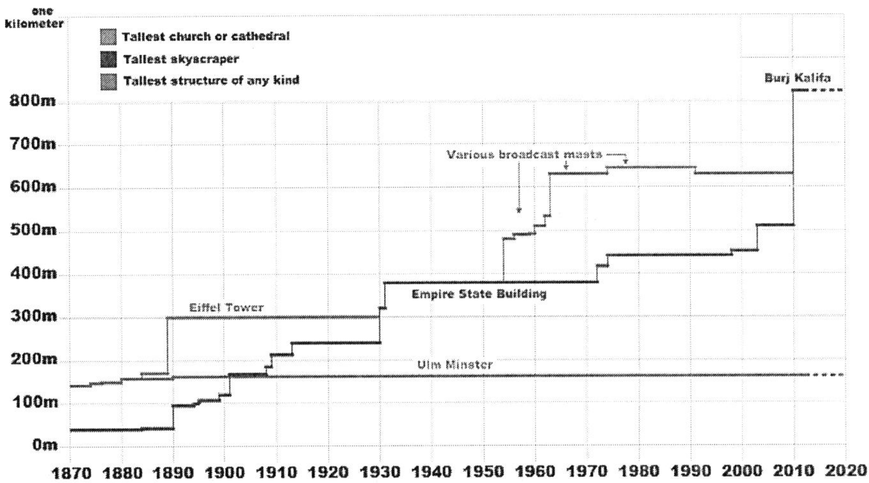

1300-present

In this chart, time progresses from right to left. Note the early buildings that lost the title as their spires collapsed.

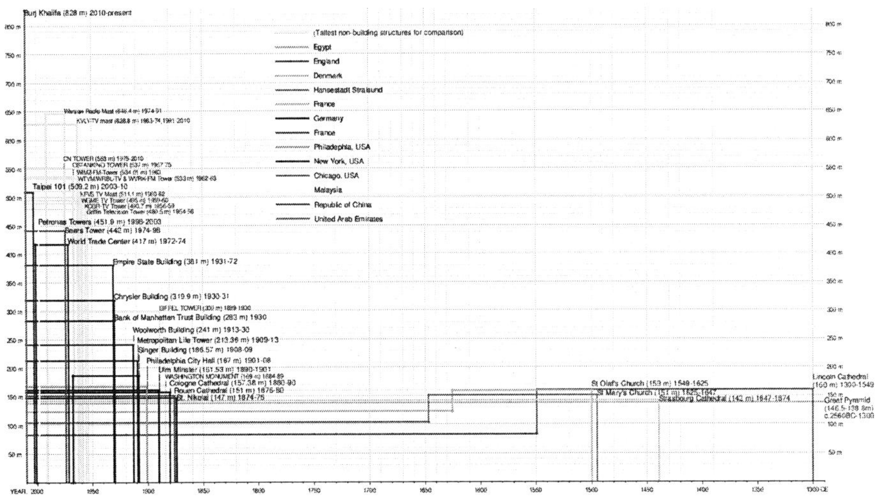

History of Increment in Height of Skyscrapers

After the construction of the Home Insurance Building in 19th century, the incrementation in the height of skyscrapers began with the construction of the Chrysler Building, followed by the Empire State Building, in New York City. The Empire State Building was the first building in the world to break the 300 metres (984 ft) barrier, and the first building to have more than 100 floors it stands at 381 metres (1,250 ft) and has 102 floors. The next tallest skyscraper was the World Trade Centre, which was completed in 1972.

The north tower was 417 m (1,368 ft) and the south 415 m (1,362 ft) tall. It surpassed the height of the Empire State Building by 36 m (118 ft). Two years later the Sears Tower was built in Chicago, standing at 442 m (1,450 ft) with 108 floors. This skyscraper surpassed the height of the World Trade Centre by 25 metres (82 ft). The Petronas Towers rose 10 metres above the Sears Tower, standing at a height of 452 m (1,483 ft) and each having 88 floors.

In 2003 the construction of Taipei 101 brought the height of skyscrapers to a new level, standing at 509 m (1,670 ft) with 101 floors. It is 59 metres (194 ft) taller than the previous record holders, the Petronas Towers. Burj Khalifa surpassed the height of Taipei 101 by 319 metres (1,047 ft) in 2009, making it 60 % taller.

It has broken several skyscraper records, and it is almost twice tall as the Empire State Building. Burj Khalifa has also broken the record of the world's tallest structure.

History of Supertall Skyscrapers by Location

Since the early skyscraper boom that took place in North America, the significant number of skyscrapers in North America have dominated the 100 tallest buildings in the world. In 1930, 99 of the tallest 100 were located in North America, with 51 percent in New York City alone. By future this percentage is expected to be dramatically decreased to only 22 percent and 5 percent respectively.

The predominance of skyscrapers in North America is rapidly decreasing due to the explosion of skyscraper construction in other parts of the world, especially in Asia.

In Asia there has been a dramatic increase in the number of supertall skyscrapers beginning with the construction of Petronas Twin Towers. There are currently sixty buildings in the world's 100 tallest that are located in Asia (including the Middle East).

Increment in Usage of Skyscrapers

Since the skyscraper boom, the great majority of skyscrapers in the world were used predominantly as office space. From 1930 to 2000 the percentage of office towers never fell below 86 percent and in future it is expected to be as low as 46 percent. In 2010 less than half of the 100 tallest buildings in the world were office towers with the majority utilised as residential and mixed use. Today, only four out of the ten tallest buildings in the world, and twenty eight out of the fifty tallest in the world, are used primarily as offices.

A mixed-use tall building contains two or more functions (uses), where each of the functions occupies a significant proportion of the tower's total space. Support areas such as car parks and mechanical plant space do not contribute towards mixed-use status.

Skyscrapers used as hotels and as residential space are on the low side there are only a few supertall skyscrapers of that type in the 100 tallest skyscrapers in the world. The tallest residential building in the world is Q1, on the Gold Coast, Australia at 323 m (1,060 ft). But in the future, several supertall residential towers will be built globally. Some notable skyscrapers are Pentominium, the Princess Tower, and Marina 101. There are few supertall hotel skyscrapers in the world. The Emirates Park Towers Hotel & Spa, Rose Tower and Burj Al Arab are the three only supertall hotels standing, while the topped out Emirates Park Towers Hotel & Spa, is currently the tallest hotel in the world at 377 m (1,237 ft).

Ground Scraper

Groundscraper is a late 20th-century neologism for a large building that is only around a dozen stories high but which greatly extends horizontally. It is thus a skyscraper that is close to the ground.

Definition

MSN Encarta defines groundscraper as:

> *"a large low or medium-rise building, typically containing offices, that spreads horizontally and occupies a large amount of land"*

Examples

- Swiss bank UBS is planning the largest office building in the City of London; the design of *5 Broadgate* has been labelled a groundscraper.

- China Vanke headquarters in Shenzhen is as large as the Empire State Building, but is laid out horizontally and 5 stories above ground level. A park occupies the space below.

Sustainable Design

Sustainable design (also called environmental design, environmentally sustainable design, environmentally conscious design, etc.) is the philosophy of designing physical objects, the built environment, and services to comply with the principles of economic, social, and ecological sustainability.

Intentions

The intention of sustainable design is to "eliminate negative environmental impact completely through skillful, sensitive design". Manifestations of sustainable design require no non-renewable resources, impact the environment minimally, and relate people with the natural environment. Beyond the "elimination of negative environmental impact", sustainable design must create projects that are meaningful innovations that can shift behaviour. A dynamic balance between economy and society, to generate long-term relationships between user and object or service and finally respectful and be mindful of the environmental and social differences.

Applications

Applications of this philosophy range from the microcosm — small objects for everyday use, through to the macrocosm — buildings, cities, and the Earth's physical surface. It is a philosophy that can be applied in the fields of architecture, landscape architecture, urban design, urban planning, engineering, graphic design, industrial design, interior design, fashion design and human-computer interaction.

Sustainable design is mostly a general reaction to global environmental crises, the rapid growth of economic activity and human population, depletion of natural resources, damage to ecosystems, and loss of biodiversity.

The limits of sustainable design are reducing. Whole earth impacts are beginning to be considered because growth in goods and services is consistently outpacing gains in efficiency. As a result, the net effect of sustainable design to date has been to simply improve the efficiency of rapidly increasing impacts. The present approach, which focuses on the efficiency of delivering individual goods and services, does not solve this problem. The basic dilemmas include: the increasing complexity of efficiency improvements; the difficulty of implementing new technologies in societies built around old ones; that physical impacts of delivering goods and services are not localized, but are distributed throughout the economies; and that the scale of resource use is growing and not stabilizing.

Sustainable Design Principles

While the practical application varies among disciplines, some common principles are as follows:

- Low-impact materials: choose non-toxic, sustainably produced or recycled materials which require little energy to process
- Energy efficiency: use manufacturing processes and produce products which require less energy
- Quality and durability: longer-lasting and better-functioning products will have to be replaced less frequently, reducing the impacts of producing replacements
- Design for reuse and recycling: "Products, processes, and systems should be designed for performance in a commercial 'afterlife'."
- Design Impact Measures for total carbon footprint and life-cycle assessment for any resource used are increasingly required and available. Many are complex, but some give quick and accurate whole-earth estimates of impacts. One measure estimates any spending as consuming an average economic share of global energy use of 8,000 BTU (8,400 kJ) per dollar and producing CO2 at the average rate of 0.57 kg of CO2 per dollar (1995 dollars US) from DOE figures.
- Sustainable Design Standards and project design guides are also increasingly available and are vigorously being developed by a wide array of private organisations and individuals. There is also a large body of new methods emerging from the rapid development of what has become known as 'sustainability science' promoted by a wide variety of educational and governmental institutions.
- Biomimicry: "redesigning industrial systems on biological lines ... enabling the constant reuse of materials in continuous closed cycles..."
- Service substitution: shifting the mode of consumption from personal ownership of products to provision of services which provide similar functions, e.g., from a private automobile to a carsharing service. Such a system promotes minimal resource

use per unit of consumption (e.g., per trip driven).

- Renewability: materials should come from nearby (local or bioregional), sustainably managed renewable sources that can be composted when their usefulness has been exhausted.
- Robust eco-design: robust design principles are applied to the design of a pollution sources).

Bill of Rights for the Planet

A model of the new design principles necessary for sustainability is exemplified by the "Bill of Rights for the Planet" or "Hannover Principles" - developed by William McDonough Architects for EXPO 2000 that was held in Hannover, Germany.

The Bill of Rights:

1. Insist on the right of humanity and nature to co-exist in a healthy, supportive, diverse, and sustainable condition.
2. Recognise Interdependence. The elements of human design interact with and depend on the natural world, with broad and diverse implications at every scale. Expand design considerations to recognising even distant effects.
3. Respect relationships between spirit and matter. Consider all aspects of human settlement including community, dwelling, industry, and trade in terms of existing and evolving connections between spiritual and material consciousness.
4. Accept responsibility for the consequences of design decisions upon human well-being, the viability of natural systems, and their right to co-exist.
5. Create safe objects of long-term value. Do not burden future generations with requirements for maintenance or vigilant administration of potential danger due to the careless creations of products, processes, or standards.
6. Eliminate the concept of waste. Evaluate and optimize the full life-cycle of products and processes, to approach the state of natural systems in which there is no waste.
7. Rely on natural energy flows. Human designs should, like the living world, derive their creative forces from perpetual solar income. Incorporate this energy efficiently and safely for responsible use.
8. Understand the limitations of design. No human creation lasts

forever and design does not solve all problems. Those who create and plan should practice humility in the face of nature. Treat nature as a model and mentor, not an inconvenience to be evaded or controlled.

9. Seek constant improvement by the sharing of knowledge. Encourage direct and open communication between colleagues, patrons, manufacturers and users to link long term sustainable considerations with ethical responsibility, and re-establish the integral relationship between natural processes and human activity.

These principles were adopted by the World Congress of the International Union of Architects (UIA) in June 1993 at the American Institute of Architects' (AIA) Expo 93 in Chicago. Further, the AIA and UIA signed a "Declaration of Interdependence for a Sustainable Future."

In summary, the declaration states that today's society is degrading its environment and that the AIA, UIA, and their members are committed to:

- Placing environmental and social sustainability at the core of practices and professional responsibilities
- Developing and continually improving practices, procedures, products, services, and standards for sustainable design
- Educating the building industry, clients, and the general public about the importance of sustainable design
- Working to change policies, regulations, and standards in government and business so that sustainable design will become the fully supported standard practice
- Bringing the existing built environment up to sustainable design standards

In addition, the Interprofessional Council on Environmental Design (ICED), a coalition of architectural, landscape architectural, and engineering organisations, developed a vision statement in an attempt to foster a team approach to sustainable design.

ICED states: The ethics, education and practices of our professions will be directed to shape a sustainable future. . . . To achieve this vision we will join . . . as a multidisciplinary partnership."

These activities are an indication that the concept of sustainable design is being supported on a global and interprofessional scale and that the ultimate goal is to become more environmentally responsive.

The world needs facilities that are more energy efficient and that promote conservation and recycling of natural and economic resources.

Conceptual Problems to Solve

- Diminishing Returns: The principle that all directions of progress run out, ending with diminishing returns, is evident in the typical 'S' curve of The Technology Life Cycle and in the useful life of any system as discussed in Industrial Ecology and Life Cycle Assessment. It's as reliable an expectation as any principle of science that diminishing returns signal natural limits. Common office and business management practice is to read diminishing returns in any direction of effort as an indication of diminishing opportunity, a potential for accelerating their decline and signal to turn elsewhere.
- Unsustainable Investment: A problem arises when the limits of a resource are hard to see, so increasing investment in response to diminishing returns may seem profitable as in the Tragedy of the Commons, but may lead to a collapse. This problem of increasing investment in diminishing resources has also been studied in relation to the causes of civilization collapse by Joseph Tainter among others. This natural error in investment policy contributed to the collapse of both the Roman and Mayan, among others. Relieving over-stressed resources requires reducing pressure on them, not continually increasing it whether more efficiently or not

Waste Prevention

Negative Effects of Waste

About 80 million tonnes of waste in total are generated in the U.K. alone, for example, each year. And with reference to only household waste, between 1991/92 and 2007/08, each person in Engerland generated an average of 1.35 pounds of waste per day.

Experience has now shown that there is no completely safe method of waste disposal. All forms of disposal have negative impacts on the environment, public health, and local economies. Landfills have contaminated drinking water. Garbage burned in incinerators has poisoned air, soil, and water. The majority of water treatment systems change the local ecology. Attempts to control or manage wastes after they are produced fail to eliminate environmental impacts.

The toxic components of household products pose serious health risks and aggravate the trash problem. In the U.S., about eight pounds in every ton of household garbage contains toxic materials, such as lead, cadmium, and mercury from batteries, insect sprays, nail polish, cleaners, and other products. When burned or buried, toxic materials also pose a serious threat to public health and the environment.

The only way to avoid environmental harm from waste is to prevent its generation. Pollution prevention means changing the way activities are conducted and eliminating the source of the problem. It does not mean doing without, but doing differently. For example, preventing waste pollution from litter caused by disposable beverage containers does not mean doing without beverages; it just means using refillable bottles.

Waste Prevention Strategies In planning for facilities, a comprehensive design strategy is needed for preventing generation of solid waste. A good garbage prevention strategy would require that everything brought into a facility be recycled for reuse or recycled back into the environment through biodegradation. This would mean a greater reliance on natural materials or products that are compatible with the environment.

Any resource-related development is going to have two basic sources of solid waste — materials purchased and used by the facility and those brought into the facility by visitors. The following waste prevention strategies apply to both, although different approaches will be needed for implementation:

- use products that minimise waste and are nontoxic
- compost or anaerobically digest biodegradable wastes
- reuse materials onsite or collect suitable materials for offsite recycling

Examples of Sustainable Design

Sustainable Planning

Urban planners that are interested in achieving sustainable development or sustainable cities use various design principles and techniques when designing cities and their infrastructure. These include Smart Growth theory, Transit-oriented development, sustainable urban infrastructure and New Urbanism. Smart Growth is an urban planning and transportation theory that concentrates growth in infill sites within the existing infrastructure of a city or town to avoid urban sprawl; and advocates compact, transit-oriented development, walkable, bicycle-friendly land use, including mixed-use development with a range of housing choices. Transit-oriented development attempts to maximise access to public transport and thereby reduce the need for private vehicles. Public transport is considered a form of Sustainable urban infrastructure, which is a design approach which promotes protected areas, energy-efficient buildings, wildlife corridors and distributed, rather than centralized, power generation and waste water treatment. New Urbanism is more of a social and aesthetic urban design movement than a green one, but it does emphasize diversity of land use and population, as well as walkable communities which inherently reduce the need for automotive travel. Both urban and rural planning can benefit from including sustainability as a central criterion when laying out roads, streets, buildings and other components of the built environment. Conventional planning practice often ignores or discounts the natural configuration of the land during the planning stages, potentially causing ecological damage such as the stagnation of streams, mudslides, soil erosion, flooding and pollution. Applying methods such as scientific modelling to planned building projects can draw attention to problems before construction begins, helping to minimise damage to the natural environment.

Cohousing is an approach to planning based on the idea of intentional communities. Such projects often prioritize common space over private space resulting in grouped structures that preserve more of the surrounding environment.

Watershed assessment of carrying capacity; estuary, riparian zone restoration and groundwater recharge for hydrologic cycle viability; and other opportunities and issues about Water and the environment show that the foundation of smart growth lies in the protection and preservation of water resources. The total amount of precipitation landing on the surface of a community becomes the supply for the inhabitants. This supply amount then dictates the carrying capacity - the potential population - as supported by the "water crop."

Sustainable Architecture

Sustainable architecture is the design of sustainable buildings. Sustainable architecture attempts to reduce the collective environmental impacts during the production of building components, during the construction process, as well as during the lifecycle of the building (heating, electricity use, carpet cleaning etc.) This design practice emphasizes efficiency of heating and cooling systems;

alternative energy sources such as solar hot water, appropriate building siting, reused or recycled building materials; on-site power generation - solar technology, ground source heat pumps, wind power; rainwater harvesting for gardening, washing and aquifer recharge; and on-site waste management such as green roofs that filter and control stormwater runoff. This requires close cooperation of the design team, the architects, the engineers, and the client at all project stages, from site selection, scheme formation, material selection and procurement, to project implementation.

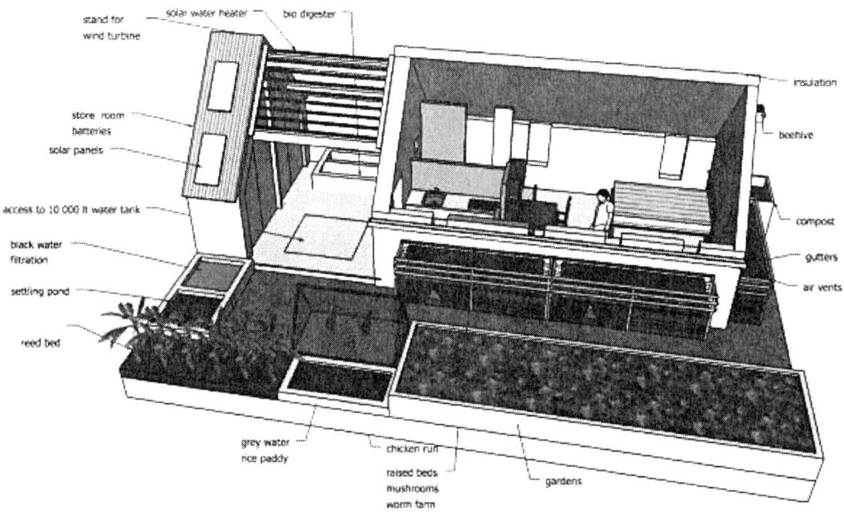

Figure : Sustainable building design

Sustainable architects design with sustainable living in mind. Sustainable vs green design is the challenge that designs not only reflect healthy processes and uses but are powered by renewable energies and site specific resources. A test for sustainable design is — can the design function for its intended use without fossil fuel — unplugged. This challenge suggests architects and planners design solutions that can function without pollution rather than just reducing pollution. As technology progresses in architecture and design theories and as examples are built and tested, architects will soon be able to create not only passive, null-emission buildings, but rather be able to integrate the entire power system into the building design. In 2004 the 59 home housing community, the Solar Settlement, and a 60,000 sq ft (5,600 m^2) integrated retail, commercial and residential building, the Sun Ship, were completed by architect Rolf Disch in

Freiburg, Germany. The Solar Settlement is the first housing community world wide in which every home, all 59, produce a positive energy balance.

An essential element of Sustainable Building Design is indoor environmental quality including air quality, illumination, thermal conditions, and acoustics. The integrated design of the indoor environment is essential and must be part of the integrated design of the entire structure. ASHRAE Guideline 10-2011 addresses the interactions among indoor environmental factors and goes beyond traditional standards.

Sustainable Landscape and Garden Design

Sustainable landscape architecture is a category of sustainable design and energy-efficient landscaping concerned with the planning and design of outdoor space. Design techniques include planting trees to shade buildings from the sun or protect them from wind, using local materials, on-site composting and chipping to reduce green waste hauling, and also may involve using drought-resistant plantings in arid areas (xeriscaping) and buying stock from local growers to avoid energy use in transportation. Areas of the garden and landscape can also be allowed to grow wild to encourage bio-diversity.

Sustainable graphic design considers the environmental impacts of graphic design products (such as packaging, printed materials, publications, etc.) throughout a life cycle that includes: raw material; transformation; manufacturing; transportation; use; and disposal. Techniques for sustainable graphic design include: reducing the amount of materials required for production; using paper and materials made with recycled, post-consumer waste; printing with low-VOC inks; and using production and distribution methods that require the least amount of transport.

Sustainable Agriculture

Sustainable agriculture adheres to three main goals:

- environmental health,
- economic profitability,
- social and economic equity.

A variety of philosophies, policies and practices have contributed to these goals. People in many different capacities, from farmers to consumers, have shared this vision and contributed to it. Despite the diversity of people and perspectives, the following themes commonly weave through definitions of sustainable agriculture.

There are strenuous discussions — among others by the agricultural sector and authorities — if existing pesticide protocols and methods of soil conservation adequately protect topsoil and wildlife. Doubt has risen if these are sustainable, and if agrarian reforms would permit an efficient agriculture with fewer pesticides, therefore reducing the damage to the ecosystem.

Domestic Machinery and Furniture

Automobiles, home appliances and furnitures can be designed for repair and disassembly (for recycling), and constructed from recyclable materials such as steel, aluminium and glass, and renewable materials, such as Zelfo, wood and plastics from natural feedstocks. Careful selection of materials and manufacturing processes can often create products comparable in price and performance to non-sustainable products. Even mild design efforts can greatly increase the sustainable content of manufactured items.

Improvements to heating, cooling, ventilation and water heating:

- Absorption refrigerator
- Annualized geothermal solar
- Earth cooling tubes
- Geothermal heat pump
- Heat recovery ventilation
- Hot water heat recycling
- Passive cooling
- Renewable heat
- Seasonal thermal storage
- Solar air conditioning
- Solar hot water.

Disposable Products

Detergents, newspapers and other disposable items can be designed to decompose, in the presence of air, water and common soil organisms. The current challenge in this area is to design such items in attractive colours, at costs as low as competing items. Since most such items end up in landfills, protected from air and water, the utility of such disposable products is debated.

Eco Fashion and Home Accessories

Creative designers and artists are perhaps the most inventive when it comes to upcycling or creating new products from old waste.

A growing number of designers upcycle waste materials such as car window glass and recycled ceramics, textile offcuts from upholstery companies, and even decommissioned fire hose to make belts and bags. Whilst accessories may seem trivial when pitted against green scientific breakthroughs; the ability of fashion and retail to influence and inspire consumer behaviour should not be underestimated. Eco design may also use bi-products of industry, reducing the amount of waste being dumped in landfill, or may harness new sustainable materials or production techniques e.g. fabric made from recycled PET plastic bottles or bamboo textiles.

Energy Sector

Sustainable technology in the energy sector is based on utilising renewable sources of energy such as solar, wind, hydro, bioenergy, geothermal, and hydrogen. Wind energy is the world's fastest growing energy source; it has been in use for centuries in Europe and more recently in the United States and other nations. Wind energy is captured through the use of wind turbines that generate and transfer electricity for utilities, homeowners and remote villages. Solar power can be harnessed through photovoltaics, concentrating solar, or solar hot water and is also a rapidly growing energy source.

The availability, potential, and feasibility of primary renewable energy resources must be analyzed early in the planning process as part of a comprehensive energy plan. The plan must justify energy demand and supply and assess the actual costs and benefits to the local, regional, and global environments. Responsible energy use is fundamental to sustainable development and a sustainable future. Energy management must balance justifiable energy demand with appropriate energy supply. The process couples energy awareness, energy conservation, and energy efficiency with the use of primary renewable energy resources.

Water Sector

Sustainable water technologies have become an important industry segment with several companies now providing important and scalable solutions to supply water in a sustainable manner.

Beyond the use of certain technologies, Sustainable Design in Water Management also consists very importantly in correct implementation of concepts. Among one of these principal concepts is the fact normally in developed countries 100% of water destined for consumption, that is not necessarily for drinking purposes, is of potable water quality. This concept of differentiating qualities of water for different purposes has been called "fit-for-purpose". This more rational use of water achieves several economies, that are not only related to water itself, but also the consumption of energy, as to achieve water of drinking quality can be extremely energy intensive for several reasons.

Sustainable Technologies

Sustainable technologies use less energy, fewer limited resources, do not deplete natural resources, do not directly or indirectly pollute the environment, and can be reused or recycled at the end of their useful life. There is a significant overlap with appropriate technology, which emphasizes the suitability of technology to the context, in particular considering the needs of people in developing countries. However, the most appropriate technology may not be the most sustainable one; and a sustainable technology may have high cost or maintenance requirements that make it unsuitable as an "appropriate technology," as that term is commonly used.

Encouraging Sustainability

The Passivhaus-Institut promotes and establishes standards for the Passive House - Passivhaus international program for Low-energy houses and other low-energy building techniques and structures.

The use of sustainable technologies may be encouraged through means such as reducing the capacity of the electrical cable supplying a home, such as Australia's Crystal Waters Village. In some cases the electricity supplier charges a higher rate for the energy used when the capacity of the supply is increased.

Terminology

In some countries the term *sustainable design* is known as Ecodesign, green design or environmental design. Victor Papanek, embraced social design and social quality and ecological quality, but did not explicitly combine these areas of design concern in one term. Over the past years the terms *sustainable design* and *design for sustainability* became more used, including the triple bottom line (people, planet and profit).

However the triple bottom line is a limited vision to describe what sustainable design and design for sustainability is. Sustainability is complicated. The decentralized nature of resources, the complexities of the issues and the lack of filtering for how they relate to design appear to be the main barriers for turning motivation into action.

"The Living principles" weave together environmental protection, social equity, and economic health — thus building upon commonly accepted, triple bottom-line frameworks. But most significantly, they incorporate cultural vitality because culture is where all aspects of sustainability find their way into the blood stream of society, and culture is where designers have the deepest impact as their creations and choices shape habits and values

Green Building

Green building (also known as green construction or sustainable building) refers to a structure and using process that is environmentally responsible and resource-efficient throughout a building's life-cycle: from siting to design, construction, operation, maintenance, renovation, and demolition. This practice expands and complements the classical building design concerns of economy, utility, durability, and comfort.

Although new technologies are constantly being developed to complement current practices in creating greener structures, the common objective is that green buildings are designed to reduce the overall impact of the built environment on human health and the natural environment by:

- Efficiently using energy, water, and other resources
- Protecting occupant health and improving employee productivity
- Reducing waste, pollution and environmental degradation

A similar concept is natural building, which is usually on a smaller scale and tends to focus on the use of natural materials that are available locally. Other related topics include sustainable design and green architecture. Sustainability may be defined as meeting the needs of present generations without compromising the ability of future generations to meet their needs. Green building does not specifically address the issue of the retrofitting existing homes.

A 2009 report by the U.S. General Services Administration found 12 sustainably designed buildings cost less to operate and have excellent energy performance. In addition, occupants were more satisfied with the overall building than those in typical commercial buildings.

Reducing Environmental Impact

Green building practices aim to reduce the environmental impact of buildings, and the very first rule is, do not build in sprawl (spreading in disordered fashion). No matter how much grass you put on your roof, no matter how many energy-efficient windows, etc., you use, if you build in sprawl, you've just defeated your purpose. Buildings account for a large amount of land. According to the National Resources Inventory, approximately 107 million acres (430,000 km^2) of land in the United States are developed. The International Energy Agency released a publication that estimated that existing buildings are responsible for more than 40% of the world's total primary energy consumption and for 24% of global carbon dioxide emissions.

Goals of Green Building

Figure : The Blu Homes mkSolaire, a green building designed by Michelle Kaufmann.

The concept of sustainable development can be traced to the energy (especially fossil oil) crisis and the environment pollution concern in the 1970s. The green building movement in the U.S. originated from the need and desire for more energy efficient and environmentally friendly construction practices. There are a number of motives to building green, including environmental, economic, and social benefits. However, modern sustainability initiatives call for an integrated and synergistic design to both new construction and in the retrofitting of an existing structure. Also known as sustainable design, this approach integrates the building life-cycle with each green practice employed with a design-purpose to create a synergy amongst the practices used.

Green building brings together a vast array of practices and techniques to reduce and ultimately eliminate the impacts of buildings on the environment and human health. It often emphasizes taking advantage of renewable resources, e.g., using sunlight through passive solar, active solar, and photovoltaic techniques and using plants and trees through green roofs, rain gardens, and for reduction of rainwater run-off. Many other techniques, such as using packed gravel or permeable concrete instead of conventional concrete or asphalt to enhance replenishment of ground water, are used as well.

While the practices, or technologies, employed in green building are constantly evolving and may differ from region to region, there are fundamental principles that persist from which the method is derived: Siting and Structure Design Efficiency, Energy Efficiency, Water Efficiency, Materials Efficiency, Indoor Environmental Quality Enhancement, Operations and Maintenance Optimization, and Waste and Toxics Reduction. The essence of green building is an optimization of one or more of these principles. Also, with the proper synergistic design, individual green building technologies may work together to produce a greater cumulative effect.

On the aesthetic side of green architecture or sustainable design is the philosophy of designing a building that is in harmony with the natural features and resources surrounding the site. There are several key steps in designing sustainable buildings: specify 'green' building materials from local sources, reduce loads, optimize systems, and generate on-site renewable energy.

Life Cycle Assessment (LCA)

A life cycle assessment (LCA) can help avoid a narrow outlook on environmental, social and economic concerns by assessing a full range of impacts associated with all the stages of a process from cradle-to-grave (i.e., from extraction of raw materials through materials processing, manufacture, distribution, use, repair and maintenance, and disposal or recycling). Impacts taken into account include (among others) embodied energy, global warming potential, resource use, air pollution, water pollution, and waste.

In terms of green building, the last few years have seen a shift away from a *prescriptive* approach, which assumes that certain prescribed practices are better for the environment, toward the scientific evaluation of actual performance through LCA.

Although LCA is widely recognised as the best way to evaluate the environmental impacts of buildings (ISO 14040 provides a recognised LCA methodology), it is not yet a consistent requirement of green building rating systems and codes, despite the fact that embodied energy and other life cycle impacts are critical to the design of environmentally responsible buildings. In North America, LCA is rewarded to some extent in the Green Globes® rating system, and is part of the new American National Standard based on Green Globes, *ANSI/GBI 01-2010: Green Building Protocol for Commercial Buildings*. LCA is also included as a pilot credit in the LEED system, though a decision has not been made as to whether it will be incorporated fully into the next major revision. The state of California also included LCA as a voluntary measure in its 2010 draft *Green*

Building Standards Code.

Although LCA is often perceived as overly complex and time consuming for regular use by design professionals, research organisations such as BRE in the UK and the Athena Sustainable Materials Institute in North America are working to make it more accessible.

In the UK, the BRE *Green Guide to Specifications* offers ratings for 1,500 building materials based on LCA.

In North America, the ATHENA® *EcoCalculator for Assemblies* provides LCA results for several hundred common building assembles based on data generated by its more complex parent software, the ATHENA® *Impact Estimator for Buildings*. Athena software tools are especially useful early in the design process when material choices have far-reaching implications for overall environmental impact. They allow designers to experiment with different material mixes to achieve the most effective combination.

A more product-oriented tool is the BEES® (Building for Environmental and Economic Sustainability) software, which combines environmental measures with economic indicators to provide a final rating. Particularly useful at the specification and procurement stage of a project, BEES 4.0 includes data on 230 products (including generic and manufacturer brands) such as siding and sheathing.

Siting and Structure Design Efficiency

The foundation of any construction project is rooted in the concept and design stages. The concept stage, in fact, is one of the major steps in a project life cycle, as it has the largest impact on cost and performance. In designing environmentally optimal buildings, the objective is to minimise the total environmental impact associated with all life-cycle stages of the building project.

However, building as a process is not as streamlined as an industrial process, and varies from one building to the other, never repeating itself identically. In addition, buildings are much more complex products, composed of a multitude of materials and components each constituting various design variables to be decided at the design stage. A variation of every design variable may affect the environment during all the building's relevant life-cycle stages.

Energy Efficiency

Green buildings often include measures to reduce energy consumption – both the embodied energy required to extract, process, transport and install building materials and operating energy to provide services such as heating and power for equipment.

Figure : An eco-house at Findhorn Ecovillage with a turf roof and solar panels

As high-performance buildings use less operating energy, embodied energy has assumed much greater importance – and may make up as much as 30% of the overall life cycle energy consumption. Studies such as the U.S. LCI Database Project show buildings built primarily with wood will have a lower embodied energy than those built primarily with brick, concrete or steel.

To reduce operating energy use, high-efficiency windows and insulation in walls, ceilings, and floors increase the efficiency of the building envelope, (the barrier between conditioned and unconditioned space). Another strategy, passive solar building design, is often implemented in low-energy homes. Designers orient windows and walls and place awnings, porches, and trees to shade windows and roofs during the summer while maximising solar gain in the winter. In addition, effective window placement (daylighting) can provide more natural light and lessen the need for electric lighting during the day. Solar water heating further reduces energy costs.

Onsite generation of renewable energy through solar power, wind power, hydro power, or biomass can significantly reduce the environmental impact of the building. Power generation is generally the most expensive feature to add to a building.

Water Efficiency

Reducing water consumption and protecting water quality are key objectives in sustainable building. One critical issue of water consumption is that in many areas, the demands on the supplying aquifer exceed its ability to replenish itself. To the maximum extent feasible, facilities should increase their dependence on water that is collected, used, purified, and reused on-site. The protection and conservation of water throughout the life of a building may be accomplished by designing for dual plumbing that recycles water in toilet flushing. Waste-water may be minimised by utilising water conserving fixtures such as ultra-low flush toilets and low-flow shower heads. Bidets help eliminate the use of toilet paper, reducing sewer traffic and increasing possibilities of re-using water on-site. Point of use water treatment and heating improves both water quality and energy efficiency while reducing the amount of water in circulation. The use of non-sewage and greywater for on-site use such as site-irrigation will minimise demands on the local aquifer.

Materials Efficiency

Building materials typically considered to be 'green' include lumber from forests that have been certified to a third-party forest standard, rapidly renewable plant materials like bamboo and straw, insulating concrete forms, dimension stone, recycled stone, recycled metal, and other products that are non-toxic, reusable, renewable, and/or recyclable (e.g., Trass, Linoleum, sheep wool, panels made from paper flakes, compressed earth block, adobe, baked earth, rammed earth, clay, vermiculite, flax linen, sisal, seagrass, cork, expanded clay grains, coconut, wood fibre plates, calcium sand stone, concrete (high and ultra high performance, roman self-healing concrete) , etc.) The EPA (Environmental Protection Agency) also suggests using recycled industrial goods, such as coal combustion products, foundry sand, and demolition debris in construction projects Building materials should be extracted and manufactured locally to the building site to minimise the energy embedded in their transportation. Where possible, building elements should be manufactured off-site and delivered to site, to maximise benefits of off-site manufacture including minimising waste, maximising recycling (because manufacture is in one location), high quality elements, better OHS management, less noise and dust.

Indoor Environmental Quality Enhancement

The Indoor Environmental Quality (IEQ) category in LEED standards, one of the five environmental categories, was created to provide comfort, well-being, and productivity of occupants. The LEED IEQ category addresses design and construction guidelines especially: indoor air quality (IAQ), thermal quality, and lighting quality.

Indoor Air Quality seeks to reduce volatile organic compounds, or VOCs, and other air impurities such as microbial contaminants. Buildings rely on a properly designed ventilation system (passively/ naturally- or mechanically-powered) to provide adequate ventilation of cleaner air from outdoors or recirculated, filtered air as well as isolated operations (kitchens, dry cleaners, etc.) from other occupancies. During the design and construction process choosing construction materials and interior finish products with zero or low VOC emissions will improve IAQ. Most building materials and cleaning/maintenance products emit gases, some of them toxic, such as many VOCs including formaldehyde. These gases can have a detrimental impact on occupants' health, comfort, and productivity. Avoiding these products will increase a building's IEQ. LEED, HQE and Green Star contain specifications on use of low-emitting interior. Draft LEED 2012 is about to expand the scope of the involved products. BREEAM limits formaldehyde emissions, no other VOCs.

Also important to indoor air quality is the control of moisture accumulation (dampness) leading to mold growth and the presence of bacteria and viruses as well as dust mites and other organisms and microbiological concerns. Water intrusion through a building's envelope or water condensing on cold surfaces on the building's interior can enhance and sustain microbial growth. A well-insulated and tightly-sealed envelope will reduce moisture problems but adequate ventilation is also necessary to eliminate moisture from sources indoors including human metabolic processes, cooking, bathing, cleaning, and other activities. Personal temperature and airflow control over the HVAC system coupled with a properly designed building envelope will also aid in increasing a building's thermal quality. Creating a high performance luminous environment through the careful integration of daylight and electrical light sources will improve on the lighting quality and energy performance of a structure.

Solid wood products, particularly flooring, are often specified in environments where occupants are known to have allergies to dust or other particulates. Wood itself is considered to be hypo-allergenic and its smooth surfaces prevent the buildup of particles common in soft finishes like carpet. The Asthma and Allergy Foundation of American recommends hardwood, vinyl, linoleum tile or slate flooring instead of carpet. The use of wood products can also improve air quality by absorbing or releasing moisture in the air to moderate humidity. Interactions among all the indoor components and the occupants together form the processes that determine the indoor air quality. Extensive investigation of such processes is the subject of indoor air scientific research and is well documented in the journal Indoor Air,.

Operations and Maintenance Optimization

No matter how sustainable a building may have been in its design and construction, it can only remain so if it is operated responsibly and maintained properly. Ensuring operations and maintenance (O&M) personnel are part of the project's planning and development process will help retain the green criteria designed at the onset of the project. Every aspect of green building is integrated into the O&M phase of a building's life. The addition of new green technologies also falls on the O&M staff. Although the goal of waste reduction may be applied during the design, construction and demolition phases of a building's life-cycle, it is in the O&M phase that green practices such as recycling and air quality enhancement take place.

Waste Reduction

Green architecture also seeks to reduce waste of energy, water and materials used during construction. For example, in California nearly 60% of the state's waste comes from commercial buildings During the construction phase, one goal should be to reduce the amount of material going to landfills. Well-designed buildings also help reduce the amount of waste generated by the occupants as well, by providing on-site solutions such as compost bins to reduce matter going to landfills.

To reduce the amount of wood that goes to landfill, the CO_2 Neutral Alliance (a coalition of government, NGOs and the forest industry). The site includes a variety of resources for regulators, municipalities, developers, contractors, owner/operators and individuals/homeowners looking for information on wood recycling.

When buildings reach the end of their useful life, they are typically demolished and hauled to landfills. Deconstruction is a method of harvesting what is commonly considered "waste" and reclaiming it into useful building material. Extending the useful life of a structure also reduces waste – building materials such as wood that are light and easy to work with make renovations easier.

To reduce the impact on wells or water treatment plants, several options exist. "Greywater", wastewater from sources such as dishwashing or washing machines, can be used for subsurface irrigation, or if treated, for non-potable purposes, e.g., to flush toilets and wash cars. Rainwater collectors are used for similar purposes.

Centralized wastewater treatment systems can be costly and use a lot of energy. An alternative to this process is converting waste and wastewater into fertilizer, which avoids these costs and shows other benefits. By collecting human waste at the source and running it to a semi-centralized biogas plant with other biological waste, liquid fertilizer can be produced. This concept was demonstrated by a settlement in Lubeck Germany in the late 1990s. Practices like these provide soil with organic nutrients and create carbon sinks that remove carbon dioxide from the atmosphere, offsetting greenhouse gas emission. Producing artificial fertilizer is also more costly in energy than this process.

Cost and Payoff

The most criticized issue about constructing environmentally friendly buildings is the price. Photo-voltaics, new appliances, and modern technologies tend to cost more money. Most green buildings cost a premium of <2%, but yield 10 times as much over the entire life of the building. The stigma is between the knowledge of up-front cost vs. life-cycle cost. The savings in money come from more efficient use of utilities which result in decreased energy bills. It is projected that different sectors could save $130 Billion on energy bills. Also, higher worker or student productivity can be factored into savings and cost deductions. Studies have shown over a 20 year life period, some green buildings have yielded $53 to $71 per square foot back on investment. Confirming the rentability of green building investments, further studies of the commercial real estate market have found that LEED and Energy Star certified buildings achieve significantly higher rents, sale prices and occupancy rates as well as lower capitalization rates potentially reflecting lower investment risk.

Regulation and Operation

As a result of the increased interest in green building concepts and practices, a number of organisations have developed standards, codes and rating systems that let government regulators, building professionals and consumers embrace green building with confidence. In some cases, codes are written so local governments can adopt them as bylaws to reduce the local environmental impact of buildings.

Green building rating systems such as BREEAM (United Kingdom), LEED (United States and Canada), and CASBEE (Japan) help consumers determine a structure's level of environmental performance. They award credits for optional building features that support green design in categories such as location and maintenance of building site, conservation of water, energy, and building materials, and occupant comfort and health. The number of credits generally determines the level of achievement.

Green building codes and standards, such as the International Code Council's draft International Green Construction Code, are sets of rules created by standards development organisations that establish minimum requirements for elements of green building such as materials or heating and cooling.

Some of the major building environmental assessment tools currently in use include:

- Australia: Nabers / Green Star
- Brazil: AQUA / LEED Brasil
- Canada: LEED Canada / Green Globes / Built Green Canada
- China: GBAS
- Finland: PromisE
- France: HQE
- Germany: DGNB / CEPHEUS
- Hong Kong: HKBEAM
- India: Indian Green Building Council (IGBC) / GRIHA
- Indonesia: Green Building Council Indonesia (GBCI) / Greenship
- Italy: Protocollo Itaca / Green Building Council Italia
- Japan: CASBEE
- Korea: KGBC
- Malaysia: GBI Malaysia
- Mexico: LEED Mexico
- Netherlands: BREEAM Netherlands
- New Zealand: Green Star NZ
- Philippines: BERDE / Philippine Green Building Council
- Portugal: Lider A
- Qatar:
- Republic of China (Taiwan): Green Building Label
- Singapore: Green Mark
- South Africa: Green Star SA

- Spain: VERDE
- Switzerland: Minergie
- United States: LEED / Living Building Challenge / Green Globes / Build it Green / NAHB NGBS / International Green Construction Code (IGCC) / ENERGY STAR
- United Kingdom: BREEAM
- United Arab Emirates: Estidama
- IAPGSA Pakistan Institute of Architecture Pakistan Green Sustainable Architecture
- Jordan: EDAMA
- Czech Republic: SBToolCZ.

International Frameworks and Assessment Tools

IPCC Fourth Assessment Report

Climate Change 2007, the Fourth Assessment Report (AR4) of the United Nations Intergovernmental Panel on Climate Change (IPCC), is the fourth in a series of such reports. The IPCC was established by the World Meteorological Organisation (WMO) and the United Nations Environment Programme (UNEP) to assess scientific, technical and socio-economic information concerning climate change, its potential effects and options for adaptation and mitigation.

UNEP and Climate Change

United Nations Environment Program UNEP works to facilitate the transition to low-carbon societies, support climate proofing efforts, improve understanding of climate change science, and raise public awareness about this global challenge.

GHG Indicator

The Greenhouse Gas Indicator: UNEP Guidelines for Calculating Greenhouse Gas Emissions for Businesses and Non-Commercial Organisations

Agenda 21

Agenda 21 is a programme run by the United Nations (UN) related to sustainable development. It is a comprehensive blueprint of action to be taken globally, nationally and locally by organisations of the UN, governments, and major groups in every area in which humans impact on the environment. The number 21 refers to the 21st century.

FIDIC's PSM

The International Federation of Consulting Engineers (FIDIC) Project Sustainability Management Guidelines were created in order to assist project engineers and other stakeholders in setting sustainable development goals for their projects that are recognised and accepted by as being in the interests of society as a whole. The process is also intended to allow the alignment of project goals with local conditions and priorities and to assist those involved in managing projects to measure and verify their progress.

The Project Sustainability Management Guidelines are structured with Themes and Sub-Themes under the three main sustainability headings of Social, Environmental and Economic. For each individual Sub-Theme a core project indicator is defined along with guidance as to the relevance of that issue in the context of an individual project.

The Sustainability Reporting Framework provides guidance for organisations to use as the basis for disclosure about their sustainability performance, and also provides stakeholders a universally applicable, comparable framework in which to understand disclosed information.

The Reporting Framework contains the core product of the Sustainability Reporting Guidelines, as well as Protocols and Sector Supplements. The Guidelines are used as the basis for all reporting. They are the foundation upon which all other reporting guidance is based, and outline core content for reporting that is broadly relevant to all organisations regardless of size, sector, or location. The Guidelines contain principles and guidance as well as standard disclosures – including indicators – to outline a disclosure framework that organisations can voluntarily, flexibly, and incrementally, adopt.

Protocols underpin each indicator in the Guidelines and include definitions for key terms in the indicator, compilation methodologies, intended scope of the indicator, and other technical references.

Sector Supplements respond to the limits of a one-size-fits-all approach. Sector Supplements complement the use of the core Guidelines by capturing the unique set of sustainability issues faced by different sectors such as mining, automotive, banking, public agencies and others.

IPD Environment Code

The IPD Environment Code was launched in February 2008. The Code is intended as a good practice global standard for measuring the environmental performance of corporate buildings. Its aim is to accurately measure and manage the environmental impacts of corporate buildings and enable property executives to generate high quality, comparable performance information about their buildings anywhere in the world. The Code covers a wide range of building types (from offices to airports) and aims to inform and support the following;

- Creating an environmental strategy
- Inputting to real estate strategy
- Communicating a commitment to environmental improvement
- Creating performance targets
- Environmental improvement plans
- Performance assessment and measurement
- Life cycle assessments
- Acquisition and disposal of buildings
- Supplier management
- Information systems and data population
- Compliance with regulations
- Team and personal objectives.

IPD estimate that it will take approximately three years to gather significant data to develop a robust set of baseline data that could be used across a typical corporate estate.

ISO 21931

ISO/TS 21931:2006, Sustainability in building construction—Framework for methods of assessment for environmental performance of construction works—Part 1: Buildings, is intended to provide a general framework for improving the quality and comparability of methods for assessing the environmental performance of buildings. It identifies and describes issues to be taken into account when using methods for the assessment of environmental performance for new or existing building properties in the design, construction, operation, refurbishment and deconstruction stages. It is not an assessment system in itself but is intended be used in conjunction with, and following the principles set out in, the ISO 14000 series of standards.

Green Building by Country

- Green building in Egypt
- Green building in Germany

- Green building in Israel
- Green building in Malaysia
- Green building in Mexico
- Green building in South Africa
- Green building in the United Kingdom
- Green building in India
- Green building in the United States.

General

- Alexander Thomson, a pioneer in sustainable building
- Alternative natural materials
- Andrew Delmar Hopkins
- Arcology — high density ecological structures
- Active solar
- Autonomous building
- Building Codes Assistance Project
- Centre for Environmental Innovation in Roofing
- Deconstruction (building)
- Dimension stone
- Domotics
- Earthbag construction
- EarthCraft House
- Earthship
- Eco hotel
- Eco-building cluster (in Belgium)
- Ecohouse
- Environmental planning
- Energy-plus-house
- EnOcean
- Fab Tree Hab
- Federal Roofing Tax Credit for Energy Efficiency (in the US)
- Geo-exchange
- GovEnergy Workshop and Trade Show

- Green Building Council
- Green Home
- Green library
- Green technology
- Heat island effect
- Hot water heat recycling
- Insulating concrete form
- Leadership in Energy and Environmental Design
- List of low-energy building techniques
- Low-energy house
- Mahoney tables
- Nano House
- Natural building
- Photovoltaics
- Rainwater harvesting
- Sustainable city
- Sustainable habitat
- Sustainable House Day
- The Verifier
- Tropical green building
- Whole Building Design Guide
- World Green Building Council
- Zero-energy building.

Chapter 3

Natural Building

A natural building involves a range of building systems and materials that place major emphasis on sustainability. Ways of achieving sustainability through natural building focus on durability and the use of minimally processed, plentiful or renewable resources, as well as those that, while recycled or salvaged, produce healthy living environments and maintain indoor air quality. Natural building tends to rely on human labour, more than technology. As Michael G. Smith observes, it depends on "local ecology, geology and climate; on the character of the particular building site, and on the needs and personalities of the builders and users."

Figure : Porch of a modern timber framed home

The basis of natural building is the need to lessen the environmental impact of buildings and other supporting systems, without sacrificing comfort, health or aesthetics. To be more sustainable, natural building uses primarily abundantly available, renewable, reused or recycled materials. The use of rapidly renewable materials is increasingly a focus. In addition to relying on natural building materials, the emphasis on the architectural design is heightened. The orientation of a building, the utilisation of local climate and site conditions, the emphasis on natural ventilation through design, fundamentally lessen operational costs and positively impact the environmental. Building compactly and minimising the ecological footprint is common, as are on-site handling of energy acquisition, on-site water capture, alternate sewage treatment and water reuse.

Materials

The materials common to many types of natural building are clay and sand. When mixed with water and, usually, straw or another fibre, the mixture may form *cob* or *adobe* (clay blocks). Other materials commonly used in natural building are: earth (as rammed earth or earth bag), wood (cordwood or timber frame/post-and-beam), straw, rice-hulls, bamboo and rock. A wide variety of reused or recycled materials are common in natural building, including urbanite (salvaged chunks of used concrete), tires, tire bales, discarded bottles and other recycled glass.

Several other materials are increasingly avoided by many practitioners of this building approach, due to their major negative environmental or health impacts. These include unsustainably harvested wood, toxic wood-preservatives, portland cement-based mixes, paints and other coatings that off-gas volatile organic compounds (VOCs), and some plastics, particularly polyvinyl chloride (PVC or "vinyl") and those containing harmful plasticizers or hormone-mimicking formulations.

Techniques

Many traditional methods, techniques, and materials, are now experiencing a resurgence of popularity, however the relative popularity of these techniques differs around the World.

Adobe

One of the oldest building methods, adobe is simply clay and sand mixed with water. Sometimes chopped straw or other fibres are added for strength. The mixture is then allowed to dry in the desired shape. Usually adobe is shaped into bricks that can be stacked to form walls.

Various claims are made about the optimal proportions of clay and sand (or larger aggregate). Some say that the best adobe soil contains 15% - 30% clay to bind the material together. Others say equal proportions of clay and sand are best to prevent cracking or fragmenting of the bricks. Sometimes adobe is stabilized with a small amount of cement or asphalt emulsion to provide better weatherproofing.

The blocks can either be poured into molds and dried, or pressed into blocks. Adobe coloured with clay and polished with natural oil makes an attractive and resilient floor.

To protect the walls and reduce maintenance, adobe buildings usually have large overhanging eaves and sizeable foundations. Adobe can be plastered over with cob or lime-based mixes for both appearance and protection. Adobe has good thermal mass, meaning that it is slow to transmit heat or cold. It is not a good insulator, however, so insulation can be added (preferably on the outside), or a double wall built with airspace or insulation in between. The traditional thick, un-insulated adobe has proven to perform best in regions without harsh winters or where daily sun is predictably available during those cold periods.

Cob

The term *cob* is used to describe a monolithic building system based on a mixture of clay, sand and straw. The construction uses no forms, bricks or wooden framework; it is built from the ground up. Various forms of "mud" building have been used in many parts of the world for centuries, under a variety of names, and date from at least 10,000 years ago. Cob building began use in England prior to the 13th century, and fell out of favour after World War I, although it is seeing a resurgence today. Cob is one of the simplest and least expensive building techniques available, though it is typically very labour-intensive. Cob's other great advantage is versatility; It can easily be shaped into any form.

While cob building was falling out of favour in England by the late 19th century, thousands of cob structures have endured to the present (20,000 in Devon, England alone). It is estimated that from one third to one half of the world's population lives in earthen dwellings today. Although typically associated with "low-rise" structures, in Yemen and other Middle-Eastern countries, it has, for centuries, been used in "apartment" buildings of eight stories and more

Cob-like mixes are also used as plaster or filler in several methods of natural building, such as adobe, earth bags, timber frames, cordwood, and straw bales. Earth is thus a primary ingredient of natural building.

Cordwood

Cordwood construction is a term used for a natural building method in which "cordwood" or short lengths pieces of debarked tree are laid up crosswise with masonry or cob mixtures to build a wall. The cordwood, thus, becomes infill for the walls, usually between posts in a timber frame structure. Cordwood masonry can be combined with other methods (e.g., rammed earth, cob or light clay) to produce attractive combinations. Cordwood masonry construction provides a relatively high thermal mass, which helps to minimise fluctuations in temperature.

Earthbag

Earth is the most typical fill material used in bag-wall construction techniques. This building method utilises stacked polypropylene or natural-fibre (burlap) bags filled with earth or other mixes, with or without a stabilizer such as portland cement, to form footings, foundations, walls and even vaulted or domed roofs. In recent years, building with earth bags has become one of the increasingly practiced techniques in natural building. It facilitates self-contained, often free-form rammed-earth structures. Its growing popularity relates to its use of an abundant and readily available often site-available material (earth) in a potentially inexpensive building technique that is flexible, and easy to learn and use. However, because earth is a poor insulator, in more extreme climates other filler variations are now being explored, substituting pumice, rice-hulls or another material with better insulating value for all or part of the earth.

Rammed Earth

Rammed earth is an earth-based wall system made of compacted gravel, sand, and clay; that is extremely strong and durable. Quality rammed earth walls are dense, solid, and stone-like with great environmental benefits and superior low maintenance characteristics· As an option depending on climate or seismic concerns rigid insulation can be placed inside the wall as well as steel reinforcement. Rammed earth has been used for around 10,000 years in all types of buildings from low rise to high-rise and from small huts to palaces.

Rammed earth walls are formed in place by pounding damp sub-soil (gravel, sand, and clay) into movable, reusable forms with manual or machine-powered tampers. A mixture of around 70% aggregate (gravel, sand) and 30% clay is optimal. Cement may be added if the mix requires it or pigmentation to achieve the desired colour. Around

5-10 inches of mixed damp sub-soil are placed inside the forms and pounded to total compaction and the process is repeated until the desired height is achieved. What is left after the forms are removed is a wall that is structural and can last over 1000 years

Stone, Granite, and Concrete

Locally obtained stone has been used as natural construction material for centuries. Combined with modern engineering and materials such as concrete and steel, a durable, low-impact building can be constructed.

Straw Bale

Although grasses and straw have been in use in a range of ways in building since pre-history around the world, their incorporation in machine-manufactured modular bales seems to date back to the early 20th century in the midwestern United States, particularly the sand-hills of Nebraska, where grass was plentiful and other building materials (even quality sods) were not. Straw bale building typically consists of stacking a series of rows of bales (often in running-bond) on a raised footing or foundation, with a moisture barrier between.

Bale walls are often tied together with pins of bamboo, rebar, or wood (internal to the bales or on their faces), or with surface wire meshes, and then stuccoed or plastered, either with cementaceous mixes, lime-based formulations or earth/clay renders. Bale buildings can either have a structural frame of other materials, with bales between (simply serving as insulation and stucco substrate), referred to as "infill",or the bales may actually provide the support for openings and roof, referred to as "load-bearing" or "Nebraska-style", or a combination of framing and load-bearing may be employed, referred to a "hybrid" straw bale. Typically, bales created on farms with mobile machinery have been used ("field-bales"), but recently higher-density "recompressed" bales (or "straw-blocks") are increasing the loads that may be supported; where field bales might support around 600 pounds per linear foot of wall, the high density bales bear up to 4,000 lb./lin.ft. and more. And the basic bale-building method is now increasingly being extended to bound modules of other often-recycled materials, including tire-bales, as well as those of cardboard, paper, plastics and used carpeting, and to bag-contained "bales" of wood-chips, rice-hulls, etc.

Timber Frame

The essential elements of timber frame building—joined timbers, clay walls and thatch roofs were in place in Europe and Asia by the

9th century. It remained the common mode of house construction in northern cultures until the 19th century. Craftsmanship was, and is, an important value in timber frame building. The oldest timber frame structures (for example, the timber framed stave churches of Scandinavia) show both craftsmanship and a strong grasp of the technical aspects of structural design, as do such structures in Japan.

Figure : The completed frame of a modern timber frame home

Timber framing typically uses a "bent." A bent is a structural support, like a truss, consisting of two posts, a tie beam and two rafters. These are connected into a framework through joinery. To practice the craft, one must understand the basic structural aspects of the bent. This, along with a knowledge of joinery, are the basis of timber frame building.

Timber framing is now a modern method of construction, Ideally suited to mass house building as well as public buildings. In conjunction with a number of natural insulations and timber cladding or modern lime renders, it is possible to quickly construct a high performance, sustainable building, using completely natural products. The benefits are many—the building performs better over its lifespan, waste is reduced (much can be re-cycled, composted or used as fuel). Timber frame structures are frequently used in combination with other natural building techniques, such as cob, straw bale, or cordwood/masonry.

Related Ideas and Strategies

Other concepts, methods and strategies often (or sometimes) associated with natural building include: building "underground," earth sheltering, or berming, "green" or "living" planted roofs, thatched roofs and cement-free earthen floors, rubble-trench, or gabion foundations. To increase sustainability, various approaches to lower energy consumption are used in conjunction with natural building: sun-shading or other passive cooling techniques, passive solar heating, geo-exchange heating and cooling, "short-cycle" and "annualized" passive (and PV-assisted) solar space and water heating, hot water heat recycling, biologic air purification by indoor plants, passive or air-to-air/heat-recovery ventilation, solar or annualized cooling, insulated glazing and selective glazing films, night and cold-weather "movable" insulation, or on-site electric power generation by renewable energy in the form of photovoltaics (PV), wind generators, or micro-hydro (either with fully independent systems referred to as "off-grid" or with "grid-tied" systems feeding into the public electric network), low-voltage electric and avoidance of electro-magnetic and other possibly harmful forms of radiation.

Other green building strategies that improve conservation of resources include: rain-water catchment, storage, and purification; waste-water separation; biological waste-water purification and grey-water reuse; composting toilets, on-site snow/rain-water run-off management, bioswales, permeable paving, native or low-water-use ("xeriscape") landscapes, and accommodation of alternative-fuelled/powered and human-powered vehicles.

Timber Recycling

Timber recycling or Wood recycling is the process of turning waste timber into usable products. Recycling timber is a practice that was popularized in the early 1990s as issues such as deforestation and climate change, prompted both timber suppliers and consumers to turn to a more sustainable timber source. Recycling timber is the most environmentally friendly form of timber production and is very common in countries such as Australia and New Zealand where supplies of old wooden structures are plentiful. Timber can be chipped down into wood chips which can be used to power homes or power plants.

Benefits

Recycling timber has become popular due to its image as an environmentally friendly product, with consumers commonly believing that by purchasing recycled wood the demand for "green timber" will fall and ultimately benefit the environment. Greenpeace also view recycled timber as an environmentally friendly product, citing it as the most preferable timber source on their website. The arrival of recycled timber as a construction product has been important in both raising industry and consumer awareness towards deforestation and promoting timber mills to adopt more environmentally friendly practices.

Figure : Example of Recycled timber as a finished product

Recycling one ton of wood saves 18,000,000 BTUs of heat energy.

Drawbacks

Some hurdles facing the widespread adoption of recycled timber: Sometimes the ends of wall studs need to be trimmed off to stop decay and cracking, thus resulting in a shorter piece of wood; this trimming may result in pieces of wood that don't meet building codes. Though the price may be less than for new wood, the process of selecting usable pieces of salvaged wood, pulling out nails, and refinishing for a new use can be labourious and time-consuming.

Demolition must happen in such a way as to preserve as much of the timber as possible in a building, which means more time spent dismantling a building rather than just tearing it down quickly. The trade in recycled timber is not well-established everywhere, so a reliable supply of usable wood may be hard to come by for builders. There may be a stigma associated with using "used" or "cheap" wood that is perceived to be of not as high quality as "new" wood. Not all pieces of wood in a dismantled building will fit in a new building, and it may be cheaper and easier, from a design and labour perspective, to simply get new wood (ex: wood from a 6 foot (1.8 m) deck being used in a 7 foot (2.1 m) deck). Of course, none of these issues are insurmountable, and they are issues of convenience and logistics rather than structural integrity, but many builders find it easier and less time-consuming to simply get new wood in standard uniform sizes.

Recycling Timber

Figure : Recycled timber salvaged from a demolished building in Trat, Thailand, is applied to make a new roof.

Recycled timber most commonly comes from old buildings, bridges and wharfs, where it is carefully stripped out and put aside by demolishers. At the same time any usable dimension stone is set aside for reuse. The demolishers then sell the salvaged timber to merchants who then re-mill the timber by manually scanning it with a metal detector, which allows the timber to be de-nailed and sawn to size. Once re-milled the timber is commonly sold to consumers in the form of timber flooring, beams and decking.

Recycling

Recycling is processing used materials (waste) into new products to prevent waste of potentially useful materials, reduce the consumption of fresh raw materials, reduce energy usage, reduce air pollution (from incineration) and water pollution (from landfilling) by reducing the need for "conventional" waste disposal, and lower greenhouse gas emissions as compared to virgin production. Recycling is a key component of modern waste reduction and is the third component of the "Reduce, Reuse, Recycle" waste hierarchy.

There are some ISO standards relating to recycling such as ISO 15270:2008 for plastics waste and ISO 14001:2004 for environmental management control of recycling practice.

Recyclable materials include many kinds of glass, paper, metal, plastic, textiles, and electronics. Although similar in effect, the composting or other reuse of biodegradable waste – such as food or garden waste – is not typically considered recycling.

Materials to be recycled are either brought to a collection centre or picked up from the curbside, then sorted, cleaned, and reprocessed into new materials bound for manufacturing. In the strictest sense, recycling of a material would produce a fresh supply of the same material—for example, used office paper would be converted into new office paper, or used foamed polystyrene into new polystyrene.

However, this is often difficult or too expensive (compared with producing the same product from raw materials or other sources), so "recycling" of many products or materials involves their reuse in producing different materials (e.g., paperboard) instead. Another form of recycling is the salvage of certain materials from complex products, either due to their intrinsic value (e.g., lead from car batteries, or gold from computer components), or due to their hazardous nature (e.g., removal and reuse of mercury from various items).

Critics dispute the net economic and environmental benefits of recycling over its costs, and suggest that proponents of recycling often make matters worse and suffer from confirmation bias. Specifically, critics argue that the costs and energy used in collection and transportation detract from (and outweigh) the costs and energy saved in the production process; also that the jobs produced by the recycling industry can be a poor trade for the jobs lost in logging, mining, and other industries associated with virgin production; and that materials such as paper pulp can only be recycled a few times before material degradation prevents further recycling. Proponents of recycling dispute each of these claims, and the validity of arguments from both sides has led to enduring controversy.

History

Origins

Recycling has been a common practice for most of human history, with recorded advocates as far back as Plato in 400 BC. During periods when resources were scarce, archaeological studies of ancient waste dumps show less household waste (such as ash, broken tools and pottery)—implying more waste was being recycled in the absence of new material.

In pre-industrial times, there is evidence of scrap bronze and other metals being collected in Europe and melted down for perpetual reuse. In Britain dust and ash from wood and coal fires was collected by 'dustmen' and downcycled as a base material used in brick making. The main driver for these types of recycling was the economic advantage of obtaining recycled feedstock instead of acquiring virgin material, as well as a lack of public waste removal in ever more densely populated areas.

In 1813, Benjamin Law developed the process of turning rags into 'shoddy' and 'mungo' wool in Batley, Yorkshire. This material combined recycled fibres with virgin wool. The West Yorkshire shoddy industry in towns such as Batley and Dewsbury, lasted from the early 19th century to at least 1914.

Industrialization spurred demand for affordable materials; aside from rags, ferrous scrap metals were coveted as they were cheaper to acquire than was virgin ore. Railroads both purchased and sold scrap metal in the 19th century, and the growing steel and automobile industries purchased scrap in the early 20th century.

Many secondary goods were collected, processed, and sold by peddlers who combed dumps, city streets, and went door to door looking for discarded machinery, pots, pans, and other sources of metal. By World War I, thousands of such peddlers roamed the streets of American cities, taking advantage of market forces to recycle post-consumer materials back into industrial production.

Wartime

Resource shortages caused by the world wars, and other such world-changing occurrences greatly encouraged recycling. Massive government promotion campaigns were carried out in World War II in every country involved in the war, urging citizens to donate metals and conserve fibre, as a matter of significant patriotic importance.

For example in 1939, Britain launched the programme Paper Salvage to encourage the recycling of materials to aid the war effort. Resource conservation programs established during the war were continued in some countries without an abundance of natural resources, such as Japan, after the war ended.

Post-war

The next big investment in recycling occurred in the 1970s, due to rising energy costs. Recycling aluminium uses only 5% of the energy required by virgin production; glass, paper and metals have less dramatic but very significant energy savings when recycled feedstock is used.

Legislation

Supply

For a recycling program to work, having a large, stable supply of recyclable material is crucial. Three legislative options have been used to create such a supply: mandatory recycling collection, container deposit legislation, and refuse bans. Mandatory collection laws set recycling targets for cities to aim for, usually in the form that a certain percentage of a material must be diverted from the city's waste stream by a target date. The city is then responsible for working to meet this target.

Container deposit legislation involves offering a refund for the return of certain containers, typically glass, plastic, and metal. When a product in such a container is purchased, a small surcharge is added to the price. This surcharge can be reclaimed by the consumer if the container is returned to a collection point. These programs have been very successful, often resulting in an 80 percent recycling rate. Despite such good results, the shift in collection costs from local government to industry and consumers has created strong opposition to the creation of such programs in some areas.

A third method of increase supply of recyclates is to ban the disposal of certain materials as waste, often including used oil, old batteries, tires and garden waste. One aim of this method is to create a viable economy for proper disposal of banned products. Care must be taken that enough of these recycling services exist, or such bans simply lead to increased illegal dumping.

Government-mandated Demand

Legislation has also been used to increase and maintain a demand for recycled materials. Four methods of such legislation exist: minimum recycled content mandates, utilisation rates, procurement policies, recycled product labelling.

Both minimum recycled content mandates and utilisation rates increase demand directly by forcing manufacturers to include recycling in their operations. Content mandates specify that a certain percentage of a new product must consist of recycled material. Utilisation rates are a more flexible option: industries are permitted to meet the recycling targets at any point of their operation or even contract recycling out in exchange for [trade]able credits. Opponents to both of these methods point to the large increase in reporting requirements they impose, and claim that they rob industry of necessary flexibility.

Governments have used their own purchasing power to increase recycling demand through what are called "procurement policies." These policies are either "set-asides," which earmark a certain amount of spending solely towards recycled products, or "price preference" programs which provide a larger budget when recycled items are purchased. Additional regulations can target specific cases: in the United States, for example, the Environmental Protection Agency mandates the purchase of oil, paper, tires and building insulation from recycled or re-refined sources whenever possible.

The final government regulation towards increased demand is recycled product labelling. When producers are required to label their packaging with amount of recycled material in the product (including the packaging), consumers are better able to make educated choices. Consumers with sufficient buying power can then choose more environmentally conscious options, prompt producers to increase the amount of recycled material in their products, and indirectly increase demand. Standardized recycling labelling can also have a positive effect on supply of recyclates if the labelling includes information on how and where the product can be recycled.

Recycling Consumer Waste

Collection

A number of different systems have been implemented to collect recyclates from the general waste stream. These systems lie along the spectrum of trade-off between public convenience and government ease and expense. The three main categories of collection are "drop-off centres", "buy-back centres" and "curbside collection".

Drop-off Centres

Drop off centres require the waste producer to carry the recyclates to a central location, either an installed or mobile collection station or the reprocessing plant itself. They are the easiest type of collection to establish, but suffer from low and unpredictable throughput.

Buy-back Centres

Buy-back centres differ in that the cleaned recyclates are purchased, thus providing a clear incentive for use and creating a stable supply. The post-processed material can then be sold on, hopefully creating a profit. Unfortunately government subsidies are necessary to make buy-back centres a viable enterprise, as according to the United States National Solid Wastes Management Association it costs on average US$50 to process a ton of material, which can only be resold for US$30.

Curbside Collection

Curbside collection encompasses many subtly different systems, which differ mostly on where in the process the recyclates are sorted and cleaned. The main categories are mixed waste collection, commingled recyclables and source separation. A waste collection vehicle generally picks up the waste.

At one end of the spectrum is mixed waste collection, in which all recyclates are collected mixed in with the rest of the waste, and the desired material is then sorted out and cleaned at a central sorting facility.

This results in a large amount of recyclable waste, paper especially, being too soiled to reprocess, but has advantages as well: the city need not pay for a separate collection of recyclates and no public education is needed. Any changes to which materials are recyclable is easy to accommodate as all sorting happens in a central location.

In a Commingled or single-stream system, all recyclables for collection are mixed but kept separate from other waste. This greatly reduces the need for post-collection cleaning but does require public education on what materials are recyclable.

Source separation is the other extreme, where each material is cleaned and sorted prior to collection. This method requires the least post-collection sorting and produces the purest recyclates, but incurs additional operating costs for collection of each separate material. An extensive public education program is also required, which must be successful if recyclate contamination is to be avoided.

Source separation used to be the preferred method due to the high sorting costs incurred by commingled collection. Advances in sorting technology, however, have lowered this overhead substantially—many areas which had developed source separation programs have since switched to comingled collection.

Sorting

Once commingled recyclates are collected and delivered to a central collection facility, the different types of materials must be sorted. This is done in a series of stages, many of which involve automated processes such that a truck-load of material can be fully sorted in less than an hour. Some plants can now sort the materials automatically, known as single-stream recycling. A 30 percent increase in recycling rates has been seen in the areas where these plants exist.

Initially, the commingled recyclates are removed from the collection vehicle and placed on a conveyor belt spread out in a single layer. Large pieces of corrugated fibreboard and plastic bags are removed by hand at this stage, as they can cause later machinery to jam.

Next, automated machinery separates the recyclates by weight, splitting lighter paper and plastic from heavier glass and metal. Cardboard is removed from the mixed paper, and the most common types of plastic, PET (#1) and HDPE (#2), are collected. This separation is usually done by hand, but has become automated in some sorting centres: a spectroscopic scanner is used to differentiate between different types of paper and plastic based on the absorbed wavelengths, and subsequently divert each material into the proper collection channel.

Strong magnets are used to separate out ferrous metals, such as iron, steel, and tin-plated steel cans ("tin cans"). Non-ferrous metals are ejected by magnetic eddy currents in which a rotating magnetic field induces an electric current around the aluminium cans, which in turn creates a magnetic eddy current inside the cans. This magnetic eddy current is repulsed by a large magnetic field, and the cans are ejected from the rest of the recyclate stream.

Finally, glass must be sorted by hand based on its colour: brown, amber, green or clear.

Recycling Industrial Waste

Although many government programs are concentrated on recycling at home, a large portion of waste is generated by industry. The focus of many recycling programs done by industry is the cost-effectiveness of recycling. The ubiquitous nature of cardboard packaging makes cardboard a commonly recycled waste product by companies that deal heavily in packaged goods, like retail stores, warehouses, and distributors of goods. Other industries deal in niche or specialized products, depending on the nature of the waste materials that are present.

The glass, lumber, wood pulp, and paper manufacturers all deal directly in commonly recycled materials. However, old rubber tires may be collected and recycled by independent tire dealers for a profit.

Levels of metals recycling are generally low. In 2010, the

International Resource Panel, hosted by the United Nations Environment Programme (UNEP) published reports on metal stocks that exist within society and their recycling rates. The Panel reported that the increase in the use of metals during the 20th and into the 21st century has led to a substantial shift in metal stocks from below ground to use in applications within society above ground. For example, the in-use stock of copper in the USA grew from 73 to 238kg per capita between 1932 and 1999.

The report authors observed that, as metals are inherently recyclable, the metals stocks in society can serve as huge mines above ground. However, they found that the recycling rates of many metals are very low. The report warned that the recycling rates of some rare metals used in applications such as mobile phones, battery packs for hybrid cars and fuel cells, are so low that unless future end-of-life recycling rates are dramatically stepped up these critical metals will become unavailable for use in modern technology.

The military recycles some metals. The U.S. Navy's Ship Disposal Program uses ship breaking to reclaim the steel of old vessels. Ships may also be sunk to create an artificial reef. Uranium is a very dense metal that has qualities superior to lead and titanium for many military and industrial uses. The uranium left over from processing it into nuclear weapons and fuel for nuclear reactors is called depleted uranium, and it is used by all branches of the U.S. military use for armour-piercing shells and shielding.

The construction industry may recycle concrete and old road surface pavement, selling their waste materials for profit.

Some industries, like the renewable energy industry and solar photovoltaic technology in particular, are being proactive in setting up recycling policies even before their is considerable volume to their waste streams, anticipating future demand during their rapid growth.

Recycling Codes

In order to meet recyclers' needs while providing manufacturers a consistent, uniform system, a coding system is developed. The recycling code for plastics was introduced in 1988 by plastics industry through the Society of the Plastics Industry, Inc. Because municipal recycling programs traditionally have targeted packaging – primarily bottles and containers – the resin coding system offered a means of identifying the resin content of bottles and containers commonly found in the residential waste stream.

Cost-benefit Analysis

Table : Environmental effects of recycling

Material	*Energy savings*	*Air pollution savings*
Aluminium	95%	95%
Cardboard	24%	—
Glass	5-30%	20%
Paper	40%	73%
Plastics	70%	—
Steel	60%	—

There is some debate over whether recycling is economically efficient. Municipalities often see fiscal benefits from implementing recycling programs, largely due to the reduced landfill costs. A study conducted by the Technical University of Denmark found that in 83 percent of cases, recycling is the most efficient method to dispose of household waste. However, a 2004 assessment by the Danish Environmental Assessment Institute concluded that incineration was the most effective method for disposing of drink containers, even aluminium ones.

Fiscal efficiency is separate from economic efficiency. Economic analysis of recycling includes what economists call externalities, which are unpriced costs and benefits that accrue to individuals outside of private transactions. Examples include: decreased air pollution and greenhouse gases from incineration, reduced hazardous waste leaching from landfills, reduced energy consumption, and reduced waste and resource consumption, which leads to a reduction in environmentally damaging mining and timber activity. About 4000 minerals are known, of these only a few hundred minerals in the world are relatively common. At current rates, current known reserves of phosphorus will be depleted in the next 50 to 100 years. Without mechanisms such as taxes or subsidies to internalize externalities, businesses will ignore them despite the costs imposed on society.

To make such non-fiscal benefits economically relevant, advocates have pushed for legislative action to increase the demand for recycled materials. The United States Environmental Protection Agency (EPA) has concluded in favour of recycling, saying that recycling efforts reduced the country's carbon emissions by a net 49 million metric tonnes in 2005. In the United Kingdom, the Waste and Resources Action Programme stated that Great Britain's recycling efforts reduce CO_2 emissions by 10-15 million tonnes a year. Recycling is more efficient in densely populated areas, as there are economies of scale involved.

Certain requirements must be met for recycling to be economically feasible and environmentally effective. These include an adequate source of recyclates, a system to extract those recyclates from the waste stream, a nearby factory capable of reprocessing the recyclates, and a potential demand for the recycled products. These last two requirements are often overlooked—without both an industrial market for production using the collected materials and a consumer market for the manufactured goods, recycling is incomplete and in fact only "collection".

Many economists favour a moderate level of government intervention to provide recycling services. Economists of this mindset probably view product disposal as an externality of production and subsequently argue government is most capable of alleviating such a dilemma. However, those of the laissez faire approach to municipal recycling see product disposal as a service that consumers value. A free-market approach is more likely to suit the preferences of consumers since profit-seeking businesses have greater incentive to produce a quality product or service than does government. Moreover, economists almost always advise against government intrusion in any market with little or no externalities.

Trade in Recyclates

Certain countries trade in unprocessed recyclates. Some have complained that the ultimate fate of recyclates sold to another country is unknown and they may end up in landfills instead of reprocessed. According to one report, in America, 50–80 percent of computers destined for recycling are actually not recycled. There are reports of illegal-waste imports to China being dismantled and recycled solely for monetary gain, without consideration for workers' health or environmental damage.

Though the Chinese government has banned these practices, it has not been able to eradicate them. In 2008, the prices of recyclable waste plummeted before rebounding in 2009. Cardboard averaged about £53/tonne from 2004–2008, dropped to £19/tonne, and then went up to £59/tonne in May 2009. PET plastic averaged about £156/tonne, dropped to £75/tonne and then moved up to £195/tonne in May 2009. Certain regions have difficulty using or exporting as much of a material as they recycle. This problem is most prevalent with glass: both Britain and the U.S. import large quantities of wine bottled in green glass. Though much of this glass is sent to be recycled, outside the American Midwest there is not enough wine production to use all of the reprocessed material. The extra must be downcycled into building materials or re-inserted into the regular waste stream.

Similarly, the northwestern United States has difficulty finding markets for recycled newspaper, given the large number of pulp mills in the region as well as the proximity to Asian markets. In other areas of the U.S., however, demand for used newsprint has seen wide fluctuation.

In some U.S. states, a program called RecycleBank pays people with to recycle, receiving money from local municipalities for the reduction in landfill space which must be purchased. It uses a single stream process in which all material is automatically sorted.

Benefits and Criticism

Complete recycling is impossible from a practical standpoint. In summary, substitution and recycling strategies only delay the depletion of non-renewable stocks and therefore may buy time in the transition to true or strong sustainability, which ultimately is only guaranteed in an economy based on renewable resources

Much of the difficulty inherent in recycling comes from the fact that most products are not designed with recycling in mind. The concept of sustainable design aims to solve this problem, and was laid out in the book "*Cradle to Cradle: Remaking the Way We Make Things*" by architect William McDonough and chemist Michael Braungart. They suggest that every product (and all packaging they require) should have a complete "closed-loop" cycle mapped out for each component—a way in which every component will either return to the natural ecosystem through biodegradation or be recycled indefinitely.

As with environmental economics, care must be taken to ensure a complete view of the costs and benefits involved. For example, cardboard packaging for food products is more easily recycled than plastic, but is heavier to ship and may result in more waste from spoilage.

Energy

The amount of energy saved through recycling depends upon the material being recycled. Some, such as aluminium, save a great deal, while others may not save any. The Energy Information Administration (EIA) states on its website that "a paper mill uses 40 percent less energy to make paper from recycled paper than it does to make paper from fresh lumber." Some critics argue that it takes more energy to produce recycled products than it does to dispose of them in traditional landfill methods, since the curbside collection of recyclables often requires a second waste truck. However, recycling proponents point out that a second timber or logging truck is eliminated when paper is collected for recycling, so the net energy consumption is the same.

It is difficult to determine the exact amount of energy consumed or produced in waste disposal processes. How much energy is used in recycling depends largely on the type of material being recycled and the process used to do so. Aluminium is generally agreed to use far less energy when recycled rather than being produced from scratch. The EPA states that "recycling aluminium cans, for example, saves 95 percent of the energy required to make the same amount of aluminium from its virgin source, bauxite." In 2009 more than half of all aluminium cans produced came from recycled aluminium.

Every year, millions of tons of materials are being exploited from the earth's crust, and processed into consumer and capital goods. After decades to centuries, most of these materials are "lost". With the exception of some pieces of art or religious relics, they are no longer engaged in the consumption process. Where are they? Recycling is only an intermediate solution for such materials, although it does prolong the residence time in the anthroposphere. For thermodynamic reasons, however, recycling cannot prevent the final need for an ultimate sink.

"Every year, millions of tons of materials are being exploited from the earth's crust, and processed into consumer and capital goods. After decades to centuries, most of these materials are "lost". With the exception of some pieces of art or religious relics, they are no longer engaged in the consumption process. Where are they? Recycling is only an intermediate solution for such materials, although it does prolong the residence time in the anthroposphere. For thermodynamic reasons, however, recycling cannot prevent the final need for an ultimate sink" (Brunner, 1999).

Economist Steven Landsburg has suggested that the sole benefit of reducing landfill space is trumped by the energy needed and resulting pollution from the recycling process. Others, however, have calculated through life cycle assessment that producing recycled paper uses less energy and water than harvesting, pulping, processing, and transporting virgin trees. When less recycled paper is used, additional energy is needed to create and maintain farmed forests until these forests are as self-sustainable as virgin forests.

Other studies have shown that recycling in itself is inefficient to perform the "decoupling" of economic development from the depletion of non-renewable raw materials that is necessary for sustainable development. When global consumption of a natural resource grows by more than 1 percent per annum, its depletion is inevitable, and the best recycling can do is to delay it by a number of years. Nevertheless, if this decoupling can be achieved by other means, so that consumption of the resource is reduced below 1 percent per annum, then recycling becomes indispensable—indeed recycling rates above 80 percent are required for a significant slowdown of the resource depletion.

Costs

The amount of money actually saved through recycling depends on the efficiency of the recycling program used to do it. The Institute for Local Self-Reliance argues that the cost of recycling depends on various factors around a community that recycles, such as landfill fees and the amount of disposal that the community recycles. It states that communities start to save money when they treat recycling as a replacement for their traditional waste system rather than an add-on to it and by "redesigning their collection schedules and/or trucks."

In some cases, the cost of recyclable materials also exceeds the cost of raw materials. Virgin plastic resin costs 40 percent less than recycled resin. Additionally, a United States Environmental Protection Agency (EPA) study that tracked the price of clear glass from July 15 to August 2, 1991, found that the average cost per ton ranged from $40 to $60, while a USGS report shows that the cost per ton of raw silica sand from years 1993 to 1997 fell between $17.33 and $18.10.

In a 1996 article for *The New York Times*, John Tierney argued that it costs more money to recycle the trash of New York City than it does to dispose of it in a landfill. Tierney argued that the recycling process employs people to do the additional waste disposal, sorting, inspecting, and many fees are often charged because the processing costs used to make the end product are often more than the profit from its sale. Tierney also referenced a study conducted by the Solid Waste Association of North America (SWANA) that found in the six communities involved in the study, "all but one of the curbside recycling programs, and all the composting operations and waste-to-energy incinerators, increased the cost of waste disposal."

Tierney also points out that "the prices paid for scrap materials are a measure of their environmental value as recyclables. Scrap aluminium fetches a high price because recycling it consumes so much less energy than manufacturing new aluminium."

However, comparing the market cost of recyclable material to the cost of new raw materials ignores economic externalities - the costs that are currently not counted by the market. Creating a new piece of plastic, for instance, may cause more pollution and be less sustainable than recycling a similar piece of plastic, but these factors will not be counted in market cost. A life cycle assessment can be used to determine the levels of externalities and decide whether the recycling may be worthwhile despite unfavourable market costs. Alternatively, legal means (such as a carbon tax) can be used to bring externalities into the market, so that the market cost of the material becomes close to the true cost.

In 2003, the city of Santa Clarita, California was paying $28 per ton to put garbage into a landfill. The city then adopted a mandatory diaper recycling program that cost $1,800 per ton.

In a 2007 article, Michael Munger, the Chair of Political Science at Duke University, wrote, "... if recycling is more expensive than using new materials, it can't possibly be efficient... There is a simple test for determining whether something is a resource... or just garbage... If someone will pay you for the item, it's a resource... But if you have to pay someone to take the item away... then the item is garbage."

In a 2002 article for The Heartland Institute, Jerry Taylor, director of natural resource studies at the Cato Institute, wrote, "If it costs X to deliver newly manufactured plastic to the market, for example, but it costs 10X to deliver reused plastic to the market, we can conclude the resources required to recycle plastic are 10 times more scarce than the resources required to make plastic from scratch. And because recycling is supposed to be about the conservation of resources, mandating recycling under those circumstances will do more harm than good."

Working Conditions

The recycling of waste electrical and electronic equipment in India and China generates a significant amount of pollution. Informal recycling in an underground economy of these countries has generated an environmental and health disaster. High levels of lead (Pb), polybrominated diphenylethers (PBDEs), polychlorinated dioxins and furans, as well as polybrominated dioxins and furans (PCDD/Fs and PBDD/Fs) concentrated in the air, bottom ash, dust, soil, water and sediments in areas surrounding recycling sites. Critics also argue that while recycling may create jobs, they are often jobs with low wages and terrible working conditions. These jobs are sometimes considered to be make-work jobs that don't produce as much as the cost of wages to pay for those jobs. In areas without many environmental regulations and/or worker protections, jobs involved in recycling such as ship breaking can result in deplorable conditions for both workers and the surrounding communities

Environmental Impact

Economist Steven Landsburg, author of a paper entitled "Why I Am Not an Environmentalist," has claimed that paper recycling actually reduces tree populations. He argues that because paper companies have incentives to replenish the forests they own, large demands for paper lead to large forests. Conversely, reduced demand for paper leads to fewer "farmed" forests. Similar arguments were expressed in a 1995 article for The Free Market.

When foresting companies cut down trees, more are planted in their place. Most paper comes from pulp forests grown specifically for paper production. Many environmentalists point out, however, that "farmed" forests are inferior to virgin forests in several ways. Farmed forests are not able to fix the soil as quickly as virgin forests, causing widespread soil erosion and often requiring large amounts of fertilizer to maintain while containing little tree and wild-life biodiversity compared to virgin forests. Also, the new trees planted are not as big as the trees that were cut down, and the argument that there will be "more trees" is not compelling to forestry advocates when they are counting saplings.

Wood from tropical rainforests is rarely harvested for paper. Rainforest deforestation is mainly caused by population pressure demands for land.

Possible Income Loss and Social Costs

In some prosperous and many less prosperous countries in the world, the traditional job of recycling is performed by the entrepreneurial poor such as the karung guni, Zabaleen, the rag and bone man, waste picker, and junk man. With the creation of large recycling organisations that may be profitable, either by law or economies of scale, the poor are more likely to be driven out of the recycling and the remanufacturing market. To compensate for this loss of income to the poor, a society may need to create additional forms of societal programs to help support the poor. Like the parable of the broken window, there is a net loss to the poor and possibly the whole of a society to make recycling artificially profitable through law. However, as seen in Brazil and Argentina, waste pickers/informal recyclers are able to work alongside governments, in (semi)funded cooperatives, allowing informal recycling to be legitimized as a paying government job.

Because the social support of a country is likely less than the loss of income to the poor doing recycling, there is a greater chance that the poor will come in conflict with the large recycling organisations. This means fewer people can decide if certain waste is more economically reusable in its current form rather than being reprocessed. Contrasted to the recycling poor, the efficiency of their recycling may actually be higher for some materials because individuals have greater control over what is considered "waste."

One labour-intensive underused waste is electronic and computer waste. Because this waste may still be functional and wanted mostly by the poor, the poor may sell or use it at a greater efficiency than large recyclers.

Many recycling advocates believe that this laissez-faire individual-based recycling does not cover all of society's recycling needs. Thus, it does not negate the need for an organised recycling program. Local government often consider the activities of the recycling poor as contributing to property blight.

Public Participation in Recycling Programmes

"Between 1960 and 2000, the world production of plastic resins increased 25-fold, while recovery of the material remained below 5%." Many studies have addressed recycling behaviour and strategies to encourage community involvement in recycling programmes. It has been argued that recycling behaviour is not natural because it requires a focus and appreciation for long term planning, whereas humans have evolved to be sensitive to short term survival goals; and that to overcome this innate predisposition, the best solution would be to use social pressure to compel participation in recycling programmes. However, recent studies have concluded that social pressure is unviable in this context. One reason for this is that social pressure functions well in small group sizes of 50 to 150 individuals (common to nomadic hunter-gatherer peoples) but not in communities numbering in the millions, as we see today. Another reason is that individual recycling does not take place in the public view. In a study done by social psychologist Shawn Burn, it was found that personal contact with individuals within a neighbourhood is the most effective way to increase recycling within a community. In his study, he had 10 block leaders talk to their neighbours and convince them to recycle. A comparison group was sent fliers promoting recycling. It was found that the neighbours that were personally contacted by their block leaders recycled much more than the group without personal contact. As a result of this study, Shawn Burn believes that personal contact within a small group of people is an important factor in encouraging recycling. Another study done by Stuart Oskamp examines the effect of neighbours and friends on recycling. It was found in his studies that people who had friends and neighbours that recycled were much more likely to also recycle than those who didn't have friends and neighbours that recycled.

Garbage that is Claimed to be Recycled Actually Gets Put into Landfills Instead

In 2002, WNYC reported that 40% of the garbage that New York City residents separated for recycling actually ended up in landfills.

- Index of recycling topics
- E-Cycling (recycling of electronic components)
- Plastic recycling
- Recycling bin
- Sustainability
- Recycling (ecological)
- 2000s commodities boom.

Sustainability

Sustainability is the capacity to endure. For humans, sustainability is the long-term maintenance of well being, which has environmental, economic, and social dimensions, and encompasses the concept of union, an interdependent relationship and mutual responsible position with all living and non living things on earth. This philosophical interpretation moves well beyond definitions driven by progress oriented economic perspectives that see humans as providing stewardship, the responsible management of resource use. In ecology, sustainability describes how biological systems remain diverse and productive over time, a necessary precondition for human well-being. Long-lived and healthy wetlands and forests are examples of sustainable biological systems.

Healthy ecosystems and environments provide vital goods and services to humans and other organisms. There are two major ways of managing human impact on ecosystem services. One approach is environmental management; this approach is based largely on information gained from earth science, environmental science, and conservation biology. Another approach is management of consumption of resources, which is based largely on information gained from economics. Human sustainability interfaces with economics through the social and ecological consequences of economic activity. Moving towards sustainability is also a social challenge that entails, among other factors, international and national law, urban planning and transport, local and individual lifestyles and ethical consumerism. Ways of living more sustainably can take many forms from reorganising living conditions (e.g., ecovillages, eco-municipalities and sustainable cities), to reappraising work practices (e.g., using permaculture, green building, sustainable agriculture), or developing new technologies that reduce the consumption of resources.

Definition

The word sustainability is derived from the Latin *sustinere*. Dictionaries provide more than ten meanings for *sustain*, the main ones being to "maintain", "support", or "endure". However, since the 1980s *sustainability* has been used more in the sense of human sustainability on planet Earth and this has resulted in the most widely quoted definition of sustainability and sustainable development, that of the Brundtland Commission of the United Nations on March 20, 1987: "sustainable development is development that meets the needs of the present without compromising the ability of future generations to meet their own needs."

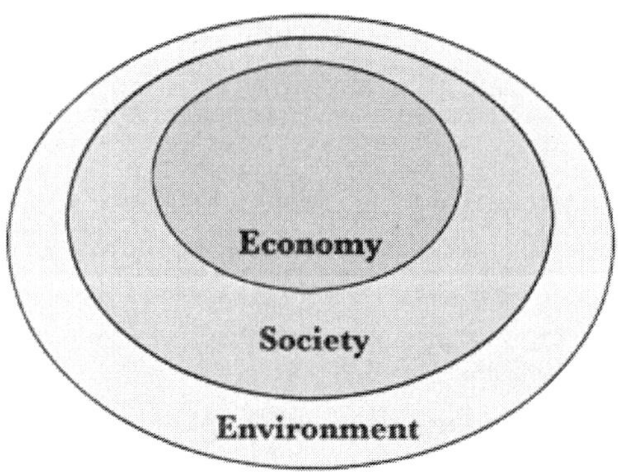

Figure : A diagram indicating the relationship between the three pillars of sustainability suggesting that both economy and society are constrained by environmental limits

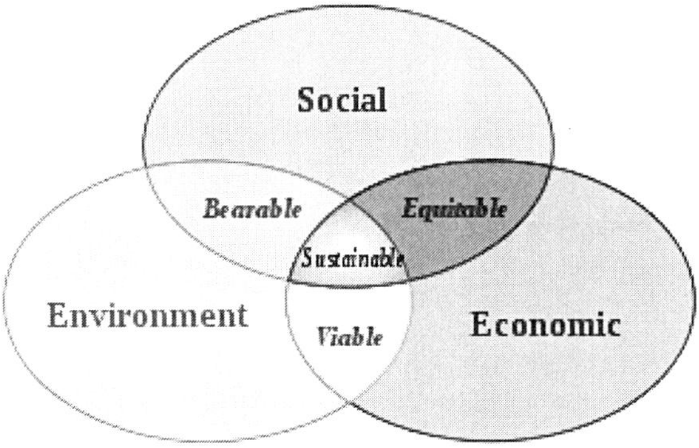

Figure : Scheme of sustainable development: at the confluence of three constituent parts.

At the 2005 World Summit it was noted that this requires the reconciliation of environmental, social and economic demands - the "three pillars" of sustainability. This view has been expressed as an illustration using three overlapping ellipses indicating that the three pillars of sustainability are not mutually exclusive and can be mutually reinforcing. The three pillars - or the 'triple bottom line' - have served as a common ground for numerous sustainability standards and certification systems in recent years, in particular in the food industry. Standards which today explicitly refer to the triple bottom line include Rainforest Alliance, Fairtrade, Utz Certified, and The Common Code for the Coffee Community. The triple bottom line is also recognised by the ISEAL Alliance-the global association for social and environmental standards.

The triple bottom line as defined by the UN is not universally accepted and has undergone various interpretations. What sustainability is, what its goals should be, and how these goals are to be achieved are all open to interpretation. For many environmentalists the idea of sustainable development is an oxymoron as development seems to entail environmental degradation. Ecological economist Herman Daly has asked, "what use is a sawmill without a forest?" From this perspective, the economy is a subsystem of human society, which is itself a subsystem of the biosphere, and a gain in one sector is a loss from another. This can be illustrated as three concentric circles.

A universally accepted definition of sustainability remains elusive because it is expected to achieve many things. On the one hand it needs to be factual and scientific, a clear statement of a specific "destination". The simple definition "sustainability is improving the quality of human life while living within the carrying capacity of supporting eco-systems", though vague, conveys the idea of sustainability having quantifiable limits. But sustainability is also a call to action, a task in progress or "journey" and therefore a political process, so some definitions set out common goals and values. The Earth Charter speaks of "a sustainable global society founded on respect for nature, universal human rights, economic justice, and a culture of peace."

To add complication the word *sustainability* is applied not only to human sustainability on Earth, but to many situations and contexts over many scales of space and time, from small local ones to the global balance of production and consumption. It can also refer to a future intention: "sustainable agriculture" is not necessarily a current situation but a goal for the future, a prediction. For all these reasons sustainability is perceived, at one extreme, as nothing more than a feel-good buzzword with little meaning or substance but, at the other, as an important but unfocused concept like "liberty" or "justice". It has also been described as a "dialogue of values that defies consensual definition".

Some researchers and institutions have pointed out that these three dimensions are not enough to reflect the complexity of contemporary society and suggest that culture could be included in this development model.

History

The history of sustainability traces human-dominated ecological systems from the earliest civilizations to the present. This history is characterised by the increased regional success of a particular society, followed by crises that were either resolved, producing sustainability, or not, leading to decline.

In early human history, the use of fire and desire for specific foods may have altered the natural composition of plant and animal communities. Between 8,000 and 10,000 years ago, Agrarian communities emerged which depended largely on their environment and the creation of a "structure of permanence."

The Western industrial revolution of the 17th to 19th centuries tapped into the vast growth potential of the energy in fossil fuels. Coal was used to power ever more efficient engines and later to generate electricity. Modern sanitation systems and advances in medicine protected large populations from disease. In the mid-20th century, a gathering environmental movement pointed out that there were environmental costs associated with the many material benefits that were now being enjoyed. In the late 20th century, environmental problems became global in scale. The 1973 and 1979 energy crises demonstrated the extent to which the global community had become dependent on non-renewable energy resources.

In the 21st century, there is increasing global awareness of the threat posed by the human-induced enhanced greenhouse effect, produced largely by forest clearing and the burning of fossil fuels.

Principles and Concepts

The philosophical and analytic framework of sustainability draws on and connects with many different disciplines and fields; in recent years an area that has come to be called sustainability science has emerged. Sustainability science is not yet an autonomous field or discipline of its own, and has tended to be problem-driven and oriented towards guiding decision-making.

Scale and Context

Sustainability is studied and managed over many scales (levels or frames of reference) of time and space and in many contexts of environmental, social and economic organisation. The focus ranges from the total carrying capacity (sustainability) of planet Earth to the sustainability of economic sectors, ecosystems, countries, municipalities, neighbourhoods, home gardens, individual lives, individual goods and services, occupations, lifestyles, behaviour patterns and so on. In short, it can entail the full compass of biological and human activity or any part of it. As Daniel Botkin, author and environmentalist, has stated: "We see a landscape that is always in flux, changing over many scales of time and space."

Consumption — Population, Technology, Resources

A major driver of human impact on Earth systems is the destruction of biophysical resources, and especially, the Earth's ecosystems. The total environmental impact of a community or of humankind as a whole depends both on population and impact per person, which in turn depends in complex ways on what resources are being used, whether or not those resources are renewable, and the scale of the human activity relative to the carrying capacity of the ecosystems involved. Careful resource management can be applied at many scales, from economic sectors like agriculture, manufacturing and industry, to work organisations, the consumption patterns of households and individuals and to the resource demands of individual goods and services. One of the initial attempts to express human impact mathematically was developed in the 1970s and is called the I PAT formula. This formulation attempts to explain human consumption in terms of three components: population numbers, levels of consumption (which it terms "affluence", although the usage is different), and impact per unit of resource use (which is termed "technology", because this impact depends on the technology used). The equation is expressed:

$$I = P \times A \times T$$

Where: I = Environmental impact, P = Population, A = Affluence, T = Technology

Measurement

Sustainability measurement is a term that denotes the measurements used as the quantitative basis for the informed management of sustainability. The metrics used for the measurement of sustainability (involving the sustainability of environmental, social and economic domains, both individually and in various combinations) are evolving: they include indicators, benchmarks, audits, sustainability standards and certification systems like Fairtrade and Organic, indexes and accounting, as well as assessment, appraisal and other reporting systems. They are applied over a wide range of spatial and temporal scales. Some of the best known and most widely used sustainability measures include corporate sustainability reporting, Triple Bottom Line accounting, World Sustainability Society and estimates of the quality of sustainability governance for individual countries using the Environmental Sustainability Index and Environmental Performance Index.

Population

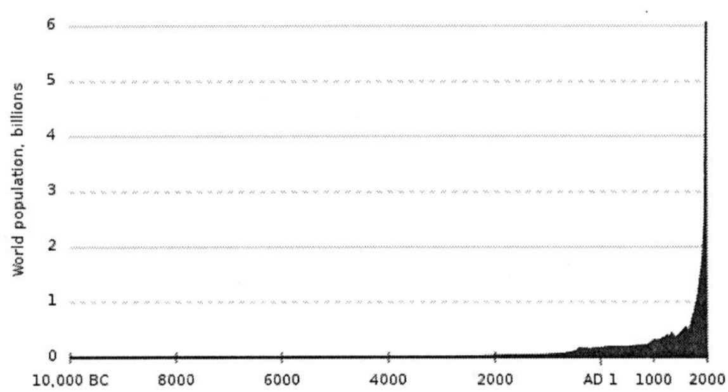

Figure : Graph showing human population growth from 10,000 BC – AD 2000,

According to the 2008 Revision of the official United Nations population estimates and projections, the world population is projected to reach 7 billion early in 2012, up from the current 6.9 billion (May 2009), to exceed 9 billion people by 2050. Most of the increase will be in developing countries whose population is projected to rise from 5.6 billion in 2009 to 7.9 billion in 2050. This increase will be distributed among the population aged 15–59 (1.2 billion) and 60 or over (1.1 billion) because the number of children under age 15 in developing countries is predicted to decrease. In contrast, the population of the more developed regions is expected to undergo only slight increase from 1.23 billion to 1.28 billion, and this would have declined to 1.15 billion but for a projected net migration from developing to developed countries, which is expected to average 2.4 million persons annually from 2009 to 2050. Long-term estimates of global population suggest a peak at around 2070 of nine to ten billion people, and then a slow decrease to 8.4 billion by 2100.

Emerging economies like those of China and India aspire to the living standards of the Western world as does the non-industrialized world in general. It is the combination of population increase in the developing world and unsustainable consumption levels in the developed world that poses a stark challenge to sustainability.

Carrying Capacity

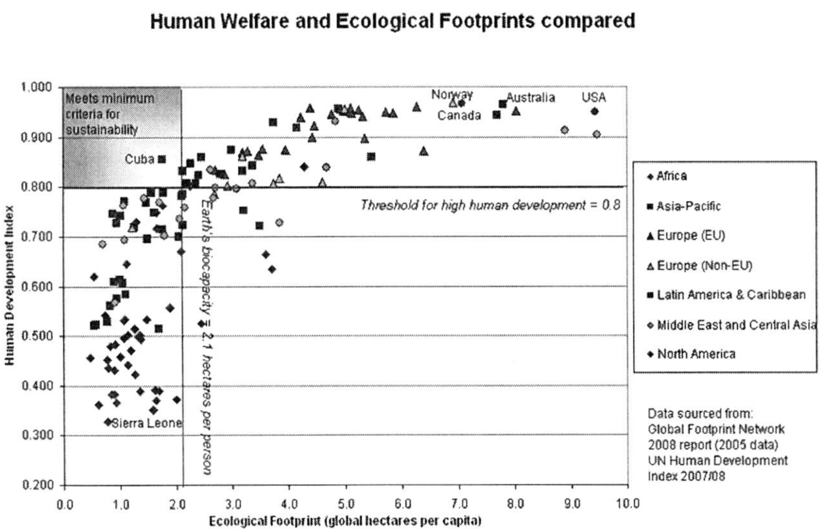

Figure : Ecological footprint for different nations compared to their Human Development Index (HDI)

At the global scale scientific data now indicates that humans are living beyond the carrying capacity of planet Earth and that this cannot continue indefinitely. This scientific evidence comes from many sources but is presented in detail in the Millennium Ecosystem Assessment and the planetary boundaries framework. An early detailed examination of global limits was published in the 1972 book *Limits to Growth*, which has prompted follow-up commentary and analysis. The Ecological footprint measures human consumption in terms of the biologically productive land needed to provide the resources, and absorb the wastes of the average global citizen. In 2008 it required 2.7 global hectares per person, 30% more than the natural biological capacity of 2.1 global hectares (assuming no provision for other organisms). The resulting ecological deficit must be met from unsustainable *extra* sources and these are obtained in three ways: embedded in the goods and services of world trade; taken from the past (e.g. fossil fuels); or borrowed from the future as unsustainable resource usage (e.g. by over exploiting forests and fisheries).

The figure (right) examines sustainability at the scale of individual countries by contrasting their Ecological Footprint with their UN Human Development Index (a measure of standard of living). The graph shows what is necessary for countries to maintain an acceptable standard of living for their citizens while, at the same time, maintaining sustainable resource use. The general trend is for higher standards of living to become less sustainable. As always, population growth has a marked influence on levels of consumption and the efficiency of resource use. The sustainability goal is to raise the global standard of living without increasing the use of resources beyond globally sustainable levels; that is, to not exceed "one planet" consumption. Information generated by reports at the national, regional and city scales confirm the global trend towards societies that are becoming less sustainable over time.

Global Human Impact on Biodiversity

At a fundamental level energy flow and biogeochemical cycling set an upper limit on the number and mass of organisms in any ecosystem. Human impacts on the Earth are demonstrated in a general way through detrimental changes in the global biogeochemical cycles of chemicals that are critical to life, most notably those of water, oxygen, carbon, nitrogen and phosphorus.

The *Millennium Ecosystem Assessment* is an international synthesis by over 1000 of the world's leading biological scientists that analyses the state of the Earth's ecosystems and provides summaries and guidelines for decision-makers. It concludes that human activity is having a significant and escalating impact on the biodiversity of world ecosystems, reducing both their resilience and biocapacity. The report refers to natural systems as humanity's "life-support system", providing essential "ecosystem services". The assessment measures 24 ecosystem services concluding that only four have shown improvement over the last 50 years, 15 are in serious decline, and five are in a precarious condition.

Environmental Dimension

Healthy ecosystems provide vital goods and services to humans and other organisms. There are two major ways of reducing negative human impact and enhancing ecosystem services and the first of these is environmental management. This direct approach is based largely on information gained from earth science, environmental science and conservation biology. However, this is management at the end of a long series of indirect causal factors that are initiated by human consumption, so a second approach is through demand management of human resource use.

Management of human consumption of resources is an indirect approach based largely on information gained from economics. Herman Daly has suggested three broad criteria for ecological sustainability: renewable resources should provide a sustainable yield (the rate of harvest should not exceed the rate of regeneration); for non-renewable resources there should be equivalent development of renewable substitutes; waste generation should not exceed the assimilative capacity of the environment.

Environmental Management

At the global scale and in the broadest sense environmental management involves the oceans, freshwater systems, land and atmosphere, but following the sustainability principle of scale it can be equally applied to any ecosystem from a tropical rainforest to a home garden.

Atmosphere

In March 2009 at a meeting of the Copenhagen Climate Council, 2,500 climate experts from 80 countries issued a keynote statement that there is now "no excuse" for failing to act on global warming and that without strong carbon reduction "abrupt or irreversible" shifts in climate may occur that "will be very difficult for contemporary societies to cope with". Management of the global atmosphere now involves assessment of all aspects of the carbon cycle to identify opportunities to address human-induced climate change and this has become a major focus of scientific research because of the potential catastrophic effects on biodiversity and human communities.

Other human impacts on the atmosphere include the air pollution in cities, the pollutants including toxic chemicals like nitrogen oxides, sulfur oxides, volatile organic compounds and particulate matter that produce photochemical smog and acid rain, and the chlorofluorocarbons that degrade the ozone layer.

Anthropogenic particulates such as sulfate aerosols in the atmosphere reduce the direct irradiance and reflectance (albedo) of the Earth's surface. Known as global dimming, the decrease is estimated to have been about 4% between 1960 and 1990 although the trend has subsequently reversed. Global dimming may have disturbed the global water cycle by reducing evaporation and rainfall in some areas. It also creates a cooling effect and this may have partially masked the effect of greenhouse gases on global warming.

Freshwater and Oceans

Water covers 71% of the Earth's surface. Of this, 97.5% is the salty water of the oceans and only 2.5% freshwater, most of which is locked up in the Antarctic ice sheet. The remaining freshwater is found in glaciers, lakes, rivers, wetlands, the soil, aquifers and atmosphere. Due to the water cycle, fresh water supply is continually replenished by precipitation, however there is still a limited amount necessitating management of this resource.

Awareness of the global importance of preserving water for ecosystem services has only recently emerged as, during the 20th century, more than half the world's wetlands have been lost along with their valuable environmental services. Increasing urbanization pollutes clean water supplies and much of the world still does not have access to clean, safe water. Greater emphasis is now being placed on the improved management of blue (harvestable) and green (soil water available for plant use) water, and this applies at all scales of water management.

Ocean circulation patterns have a strong influence on climate and weather and, in turn, the food supply of both humans and other organisms. Scientists have warned of the possibility, under the influence of climate change, of a sudden alteration in circulation patterns of ocean currents that could drastically alter the climate in some regions of the globe. Ten per cent of the world's population – about 600 million people – live in low-lying areas vulnerable to sea level rise.

Land Use

Loss of biodiversity stems largely from the habitat loss and fragmentation produced by the human appropriation of land for development, forestry and agriculture as natural capital is progressively converted to man-made capital. Land use change is fundamental to the operations of the biosphere because alterations in the relative proportions of land dedicated to urbanisation, agriculture, forest, woodland, grassland and pasture have a marked effect on the global water, carbon and nitrogen biogeochemical cycles and this can impact negatively on both natural and human systems. At the local human scale, major sustainability benefits accrue from sustainable parks and gardens and green cities.

Since the Neolithic Revolution about 47% of the world's forests have been lost to human use. Present-day forests occupy about a quarter of the world's ice-free land with about half of these occurring in the tropics. In temperate and boreal regions forest area is gradually increasing (with the exception of Siberia), but deforestation in the tropics is of major concern.

Food is essential to life. Feeding more than six billion human bodies takes a heavy toll on the Earth's resources. This begins with the appropriation of about 38% of the Earth's land surface and about 20% of its net primary productivity. Added to this are the resource-hungry activities of industrial agribusiness – everything from the crop need for irrigation water, synthetic fertilizers and pesticides to the resource costs of food packaging, transport (now a major part of global trade) and retail. Environmental problems associated with industrial agriculture and agribusiness are now being addressed through such movements as sustainable agriculture, organic farming and more sustainable business practices.

The underlying driver of direct human impacts on the environment is human consumption. This impact is reduced by not only consuming less but by also making the full cycle of production, use and disposal more sustainable. Consumption of goods and services can be analysed and managed at all scales through the chain of consumption, starting with the effects of individual lifestyle choices and spending patterns, through to the resource demands of specific goods and services, the impacts of economic sectors, through national economies to the global economy.

Analysis of consumption patterns relates resource use to the environmental, social and economic impacts at the scale or context under investigation. The ideas of embodied resource use (the total resources needed to produce a product or service), resource intensity, and resource productivity are important tools for understanding the impacts of consumption. Key resource categories relating to human needs are food, energy, materials and water.

In 2010, the International Resource Panel, hosted by the United Nations Environment Programme (UNEP), published the first global scientific assessment on the impacts of consumption and production and identified priority actions for developed and developing countries. The study found that the most critical impacts are related to ecosystem health, human health and resource depletion.

From a production perspective, it found that fossil-fuel combusting processes, agriculture and fisheries have the most important impacts. Meanwhile, from a final consumption perspective, it found that household consumption related to mobility, shelter, food and energy-using products cause the majority of life-cycle impacts of consumption.

Energy

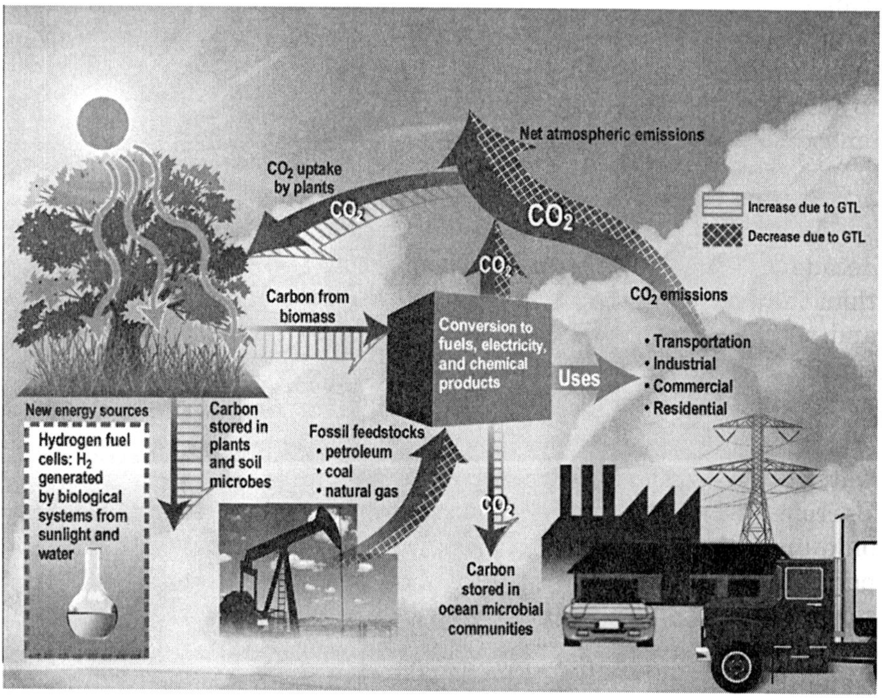

Figure : Flow of CO_2 in an ecosystem

The Sun's energy, stored by plants (primary producers) during photosynthesis, passes through the food chain to other organisms to ultimately power all living processes. Since the industrial revolution the concentrated energy of the Sun stored in fossilized plants as fossil fuels has been a major driver of technology which, in turn, has been the source of both economic and political power.

In 2007 climate scientists of the IPCC concluded that there was at least a 90% probability that atmospheric increase in CO_2 was human-induced, mostly as a result of fossil fuel emissions but, to a lesser extent from changes in land use. Stabilizing the world's climate will require high-income countries to reduce their emissions by 60–90% over 2006 levels by 2050 which should hold CO_2 levels at 450–650 ppm from current levels of about 380 ppm. Above this level, temperatures could rise by more than 2°C to produce "catastrophic" climate change. Reduction of current CO_2 levels must be achieved against a background of global population increase and developing countries aspiring to energy-intensive high consumption Western lifestyles.

Reducing greenhouse emissions, is being tackled at all scales, ranging from tracking the passage of carbon through the carbon cycle to the commercialization of renewable energy, developing less carbon-hungry technology and transport systems and attempts by individuals to lead carbon neutral lifestyles by monitoring the fossil fuel use embodied in all the goods and services they use.

Water

Water security and food security are inextricably linked. In the decade 1951–60 human water withdrawals were four times greater than the previous decade. This rapid increase resulted from scientific and technological developments impacting through the economy – especially the increase in irrigated land, growth in industrial and power sectors, and intensive dam construction on all continents.

This altered the water cycle of rivers and lakes, affected their water quality and had a significant impact on the global water cycle. Currently towards 35% of human water use is unsustainable, drawing on diminishing aquifers and reducing the flows of major rivers: this percentage is likely to increase if climate change impacts become more severe, populations increase, aquifers become progressively depleted and supplies become polluted and unsanitary. From 1961 to 2001 water demand doubled - agricultural use increased by 75%, industrial use by more than 200%, and domestic use more than 400%. In the

1990s it was estimated that humans were using 40–50% of the globally available freshwater in the approximate proportion of 70% for agriculture, 22% for industry, and 8% for domestic purposes with total use progressively increasing.

Water efficiency is being improved on a global scale by increased demand management, improved infrastructure, improved water productivity of agriculture, minimising the water intensity (embodied water) of goods and services, addressing shortages in the non-industrialised world, concentrating food production in areas of high productivity, and planning for climate change. At the local level, people are becoming more self-sufficient by harvesting rainwater and reducing use of mains water.

Food

The American Public Health Association (APHA) defines a “sustainable food system” as “one that provides healthy food to meet current food needs while maintaining healthy ecosystems that can also provide food for generations to come with minimal negative impact to the environment. A sustainable food system also encourages local production and distribution infrastructures and makes nutritious food available, accessible, and affordable to all. Further, it is humane and just, protecting farmers and other workers, consumers, and communities.” Concerns about the environmental impacts of agribusiness and the stark contrast between the obesity problems of the Western world and the poverty and food insecurity of the developing world have generated a strong movement towards healthy, sustainable eating as a major component of overall ethical consumerism. The environmental effects of different dietary patterns depend on many factors, including the proportion of animal and plant foods consumed and the method of food production.

The World Health Organisation has published a *Global Strategy on Diet, Physical Activity and Health* report which was endorsed by the May 2004 World Health Assembly. It recommends the Mediterranean diet which is associated with health and longevity and is low in meat, rich in fruits and vegetables, low in added sugar and limited salt, and low in saturated fatty acids; the traditional source of fat in the Mediterranean is olive oil, rich in monounsaturated fat. The healthy rice-based Japanese diet is also high in carbohydrates and low in fat. Both diets are low in meat and saturated fats and high in legumes and other vegetables; they are associated with a low incidence of ailments and low environmental impact.

At the global level the environmental impact of agribusiness is being addressed through sustainable agriculture and organic farming. At the local level there are various movements working towards local food production, more productive use of urban wastelands and domestic gardens including permaculture, urban horticulture, local food, slow food, sustainable gardening, and organic gardening. Sustainable seafood is seafood from either fished or farmed sources that can maintain or increase production in the future without jeopardizing the ecosystems from which it was acquired. The sustainable seafood movement has gained momentum as more people become aware about both overfishing and environmentally-destructive fishing methods.

Materials, Toxic Substances, Waste

Figure : An electric wire reel reused as a centre table in a Rio de Janeiro decoration fair. The reuse of materials is a sustainable practice that is rapidly growing among designers in Brazil.

As global population and affluence has increased, so has the use of various materials increased in volume, diversity and distance transported. Included here are raw materials, minerals, synthetic chemicals (including hazardous substances), manufactured products, food, living organisms and waste. By 2050, humanity could devour an estimated 140 billion tons of minerals, ores, fossil fuels and biomass per year three times its current appetite unless the economic growth rate is decoupled from the rate of natural resource consumption. Developed countries citizens consume an average of 16 tons of those four key resources per capita (ranging up to 40 or more tons per person in some developed countries with resource consumption levels far beyond what is likely sustainable.

Sustainable use of materials has targeted the idea of dematerialization, converting the linear path of materials (extraction, use, disposal in landfill) to a circular material flow that reuses materials as much as possible, much like the cycling and reuse of waste in nature. This approach is supported by product stewardship and the increasing use of material flow analysis at all levels, especially individual countries and the global economy.

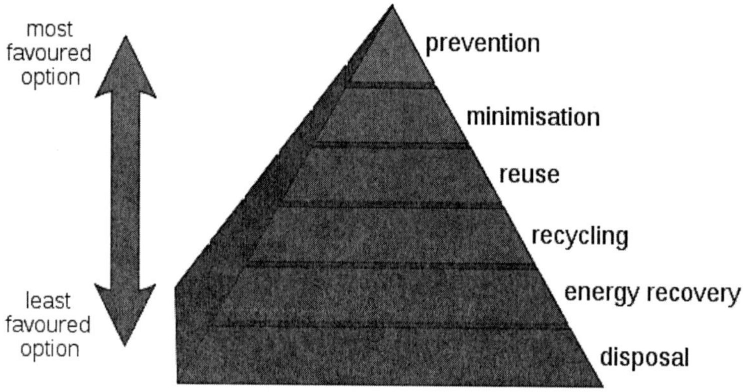

Figure : The waste hierarchy

Synthetic chemical production has escalated following the stimulus it received during the second World War. Chemical production includes everything from herbicides, pesticides and fertilizers to domestic chemicals and hazardous substances. Apart from the build-up of greenhouse gas emissions in the atmosphere, chemicals of particular concern include: heavy metals, nuclear waste, chlorofluorocarbons, persistent organic pollutants and all harmful chemicals capable of bioaccumulation. Although most synthetic chemicals are harmless there needs to be rigorous testing of new chemicals, in all countries, for adverse environmental and health effects. International legislation has been established to deal with the global distribution and management of dangerous goods.

Every economic activity produces material that can be classified as waste. To reduce waste industry, business and government are now mimicking nature by turning the waste produced by industrial metabolism into resource. Dematerialization is being encouraged through the ideas of industrial ecology, ecodesign and ecolabelling. In addition to the well-established "reduce, reuse and recycle," shoppers are using their purchasing power for ethical consumerism.

On one account, sustainability "concerns the specification of a set of actions to be taken by present persons that will not diminish the prospects of future persons to enjoy levels of consumption, wealth, utility, or welfare comparable to those enjoyed by present persons." Sustainability interfaces with economics through the social and ecological consequences of economic activity. Sustainability economics represents: "... a broad interpretation of ecological economics where environmental and ecological variables and issues are basic but part of a multidimensional perspective. Social, cultural, health-related and monetary/financial aspects have to be integrated into the analysis." However, the concept of sustainability is much broader than the concepts of sustained yield of welfare, resources, or profit margins. At present, the average per capita consumption of people in the developing world is sustainable but population numbers are increasing and individuals are aspiring to high-consumption Western lifestyles. The developed world population is only increasing slightly but consumption levels are unsustainable.

The challenge for sustainability is to curb and manage Western consumption while raising the standard of living of the developing world without increasing its resource use and environmental impact. This must be done by using strategies and technology that break the link between, on the one hand, economic growth and on the other, environmental damage and resource depletion.

A recent UNEP report proposes a green economy defined as one that "improves human well-being and social equity, while significantly reducing environmental risks and ecological scarcities": it "does not favour one political perspective over another but works to minimise excessive depletion of natural capital".

The report makes three key findings: "that greening not only generates increases in wealth, in particular a gain in ecological commons or natural capital, but also (over a period of six years) produces a higher rate of GDP growth"; that there is "an inextricable link between poverty eradication and better maintenance and conservation of the ecological commons, arising from the benefit flows from natural capital that are received directly by the poor"; "in the transition to a green economy, new jobs are created, which in time exceed the losses in "brown economy" jobs. However, there is a period of job losses in transition, which requires investment in re-skilling and re-educating the workforce".

Several key areas have been targeted for economic analysis and reform: the environmental effects of unconstrained economic growth; the consequences of nature being treated as an economic externality; and the possibility of an economics that takes greater account of the social and environmental consequences of market behaviour.

Decoupling Environmental Degradation and Economic Growth

Between economic growth and environmental degradation: as communities grow, so the environment declines. This trend is clearly demonstrated on graphs of human population numbers, economic growth, and environmental indicators. Unsustainable economic growth has been starkly compared to the malignant growth of a cancer because it eats away at the Earth's ecosystem services which are its life-support system. There is concern that, unless resource use is checked, modern global civilization will follow the path of ancient civilizations that collapsed through overexploitation of their resource base. While conventional economics is concerned largely with economic growth and the efficient allocation of resources, ecological economics has the explicit goal of sustainable scale (rather than continual growth), fair distribution and efficient allocation, in that order. The World Business Council for Sustainable Development states that "business cannot succeed in societies that fail".

In economic and environmental fields, the term decoupling is becoming increasingly used in the context of economic production and environmental quality. When used in this way, it refers to the ability of an economy to grow without incurring corresponding increases in environmental pressure. Ecological economics includes the study of societal metabolism, the throughput of resources that enter and exit the economic system in relation to environmental quality. An economy that is able to sustain GDP growth without having a negative impact on the environment is said to be decoupled.

Exactly how, if, or to what extent this can be achieved is a subject of much debate. In 2011 the International Resource Panel, hosted by the United Nations Environment Programme (UNEP), warned that by 2050 the human race could be devouring 140 billion tons of minerals, ores, fossil fuels and biomass per year – three times its current rate of consumption–unless nations can make serious attempts at decoupling.

The report noted that citizens of developed countries consume an average of 16 tons of those four key resources per capita per annum (ranging up to 40 or more tons per person in some developed countries). By comparison, the average person in India today consumes four tons per year. Sustainability studies analyse ways to reduce resource intensity (the amount of resource (e.g. water, energy, or materials) needed for the production, consumption and disposal of a unit of good or service) whether this be achieved from improved economic management, product design, or new technology.

Nature as an Economic Externality

The economic importance of nature is indicated by the use of the expression ecosystem services to highlight the market relevance of an increasingly scarce natural world that can no longer be regarded as both unlimited and free. In general, as a commodity or service becomes more scarce the price increases and this acts as a restraint that encourages frugality, technical innovation and alternative products. However, this only applies when the product or service falls within the market system. As ecosystem services are generally treated as economic externalities they are unpriced and therefore overused and degraded, a situation sometimes referred to as the Tragedy of the Commons.

One approach to this dilemma has been the attempt to "internalise" these "externalities" by using market strategies like ecotaxes and incentives, tradeable permits for carbon, and the encouragement of payment for ecosystem services. Community currencies associated with Local Exchange Trading Systems (LETS), a gift economy and Time Banking have also been promoted as a way of supporting local economies and the environment. Green economics is another market-based attempt to address issues of equity and the environment. The global recession and a range of associated government policies are likely to bring the biggest annual fall in the world's carbon dioxide emissions in 40 years.

Economic Opportunity

Treating the environment as an externality may generate short-term profit at the expense of sustainability. Sustainable business practices, on the other hand, integrate ecological concerns with social and economic ones (i.e., the triple bottom line). Growth that depletes ecosystem services is sometimes termed "uneconomic growth" as it leads to a decline in quality of life. Minimising such growth can provide opportunities for local businesses. For example, industrial waste can be treated as an "economic resource in the wrong place". The benefits of waste reduction include savings from disposal costs, fewer environmental penalties, and reduced liability insurance. This may lead to increased market share due to an improved public image. Energy efficiency can also increase profits by reducing costs.

The idea of sustainability as a business opportunity has led to the formation of organisations such as the Sustainability Consortium of the Society for Organisational Learning, the Sustainable Business Institute, and the World Council for Sustainable Development.

Research focusing on progressive corporate leaders who have embedded sustainability into commercial strategy has yielded a leadership competency model for sustainability. The expansion of sustainable business opportunities can contribute to job creation through the introduction of green-collar workers.

Social Dimension

Sustainability issues are generally expressed in scientific and environmental terms, as well as in ethical terms of stewardship, but implementing change is a social challenge that entails, among other things, international and national law, urban planning and transport, local and individual lifestyles and ethical consumerism. "The relationship between human rights and human development, corporate power and environmental justice, global poverty and citizen action, suggest that responsible global citizenship is an inescapable element of what may at first glance seem to be simply matters of personal consumer and moral choice."

Peace, Security, Social Justice

Social disruptions like war, crime and corruption divert resources from areas of greatest human need, damage the capacity of societies to plan for the future, and generally threaten human well-being and the environment. Broad-based strategies for more sustainable social systems include: improved education and the political empowerment of women, especially in developing countries; greater regard for social justice, notably equity between rich and poor both within and between countries; and intergenerational equity. Depletion of natural resources including fresh water increases the likelihood of "resource wars". This aspect of sustainability has been referred to as environmental security and creates a clear need for global environmental agreements to manage resources such as aquifers and rivers which span political boundaries, and to protect shared global systems including oceans and the atmosphere.

Sustainability and Poverty

A major hurdle to achieve sustainability is the alleviation of poverty. It has been widely acknowledged that poverty is one source of environmental degradation. Such acknowledgment has been made by the Brundtland Commission report Our Common Future and the Millennium Development Goals. According to the Brundtland report, "poverty is a major cause and effect of global environmental problems. It is therefore futile to attempt to deal with environmental problems without a broader perspective that encompasses the factors underlying world poverty and international inequality." Individuals living in poverty tend to rely heavily on their local ecosystem as a source for basic needs (such as nutrition and medicine) and general well-being. As population growth continues to increase, increasing pressure is being placed on the local ecosystem to provide these basic essentials. According to the UN Population Fund, high fertility and poverty have been strongly correlated, and the world's poorest countries also have the highest fertility and population growth rates.

The word sustainability is also used widely by western country development agencies and international charities to focus their poverty alleviation efforts in ways that can be sustained by the local populous and its environment. For example, teaching water treatment to the poor by boiling their water with charcoal, would not generally be considered a sustainable strategy, whereas using PETsolar water disinfection would be. Also, sustainable best practices can involve the recycling of materials, such as the use of recycled plastics for lumber where deforestation has devastated a countries timber base. Another example of sustainable practices in poverty alleviation is the use of exported recycled materials from developed to developing countries, such as Bridges to Prosperity's use of wire rope from shipping container gantry cranes to act as the structural wire rope for footbridges that cross rivers in poor rural areas in Asia and Africa.

Human Relationship to Nature

According to Murray Bookchin, the idea that humans must dominate nature is common in hierarchical societies. Bookchin contends that capitalism and market relationships, if unchecked, have the capacity to reduce the planet to a mere resource to be exploited. Nature is thus treated as a commodity: "The plundering of the human spirit by the market place is paralleled by the plundering of the earth by capital."

Klein argues that nature should not be alienated from man and vice versa. Hence, in order to build and maintain a balance between the two, conservation approaches need to adopt a bottom-up and community involvement approach as opposed to a top-down and Yellowstone approach which alienates man from nature based on the premise that man is harmful to nature.

Still more basically, Bookchin argued that most of the activities that consume energy and destroy the environment are senseless because they contribute little to quality of life and well being. The function of work is to legitimize, even create, hierarchy. For this reason understanding the transformation of organic into hierarchical societies is crucial to finding a way forward.

Social ecology, founded by Bookchin, is based on the conviction that nearly all of humanity's present ecological problems originate in, indeed are mere symptoms of, dysfunctional social arrangements. Whereas most authors proceed as if our ecological problems can be fixed by implementing recommendations which stem from physical, biological, economic etc., studies, Bookchin's claim is that these problems can only be resolved by understanding the underlying social processes and intervening in those processes by applying the concepts and methods of the social sciences.

Deep ecology establishes principles for the well-being of all life on Earth and the richness and diversity of life forms. This requires a substantial decrease in human population and consumption along with the reduction of human interference with the nonhuman world. To achieve this, deep ecologists advocate policies for basic economic, technological, and ideological structures that will improve the *quality of life* rather than the *standard of living*. Those who subscribe to these principles are obliged to make the necessary change happen.

Human Settlements

Sustainability principles:

1. Reduce dependence upon fossil fuels, underground metals, and minerals
2. Reduce dependence upon synthetic chemicals and other unnatural substances
3. Reduce encroachment upon nature
4. Meet human needs fairly & efficiently.

One approach to sustainable living, exemplified by small-scale urban transition towns and rural ecovillages, seeks to create self-reliant communities based on principles of simple living, which maximise self-sufficiency particularly in food production. These principles, on a broader scale, underpin the concept of a bioregional economy. Other approaches, loosely based around new urbanism, are successfully reducing environmental impacts by altering the built environment to create and preserve sustainable cities which support sustainable transport. Residents in compact urban neighbourhoods drive fewer miles, and have significantly lower environmental impacts across a range of measures, compared with those living in sprawling suburbs. Large scale social movements can influence both community choices and the built environment. Eco-municipalities may be one such movement. Eco-municipalities take a systems approach, based on sustainability principles. The eco-municipality movement is participatory, involving community members in a bottom-up approach. In Sweden, more than 70 cities and towns—25 per cent of all municipalities in the country—have adopted a common set of "Sustainability Principles" and implemented these systematically throughout their municipal operations. There are now twelve eco-municipalities in the United States and the American Planning Association has adopted sustainability objectives based on the same principles.

There is a wealth of advice available to individuals wishing to reduce their personal impact on the environment through small, inexpensive and easily achievable steps. But the transition required to reduce global human consumption to within sustainable limits involves much larger changes, at all levels and contexts of society. The United Nations has recognised the central role of education, and have declared a decade of education for sustainable development, 2005–2014, which aims to "challenge us all to adopt new behaviours and practices to secure our future". The Worldwide Fund for Nature proposes a strategy for sustainability that goes beyond education to tackle underlying individualistic and materialistic societal values head-on and strengthen people's connections with the natural world.

- Appropriate technology
- Chemical Leasing
- Conservation biology
- Cradle-to-cradle design
- Environmental issue
- Extinction
- Introduced species
- List of sustainability topics
- Micro-sustainability
- Outline of sustainability
- Permaculture
- Sociocultural evolution
- Stewardship
- Sustainability and systemic change resistance

- Sustainable development
- Sustainability standards and certification
- The Venus Project
- The Zeitgeist Movement
- World Cities Summit.

Extinction

In biology and ecology, extinction is the end of an organism or of a group of organisms (taxon), normally a species. The moment of extinction is generally considered to be the death of the last individual of the species, although the capacity to breed and recover may have been lost before this point. Because a species' potential range may be very large, determining this moment is difficult, and is usually done retrospectively. This difficulty leads to phenomena such as Lazarus taxa, where a species presumed extinct abruptly "re-appears" (typically in the fossil record) after a period of apparent absence.

Through evolution, new species arise through the process of speciation—where new varieties of organisms arise and thrive when they are able to find and exploit an ecological niche—and species become extinct when they are no longer able to survive in changing conditions or against superior competition.

The relationship between animals and their ecological niches has been firmly established. A typical species becomes extinct within 10 million years of its first appearance, although some species, called living fossils, survive virtually unchanged for hundreds of millions of years. Most extinctions have occurred naturally, prior to *Homo sapiens* walking on Earth: it is estimated that 99.9% of all species that have ever existed are now extinct. Mass extinctions are relatively rare events; however, isolated extinctions are quite common. Only recently have extinctions been recorded and scientists have become alarmed at the high rates of recent extinctions. Most species that become extinct are never scientifically documented. Some scientists estimate that up to half of presently existing species may become extinct by 2100. It is difficult to estimate the trajectory that biodiversity might have taken without human impact but scientists at the University of Bristol estimate that biodiversity might increase exponentially without human influence.

Definition

A species becomes extinct when the last existing member of that species dies. Extinction therefore becomes a certainty when there are no surviving individuals that are able to reproduce and create a new generation. A species may become functionally extinct when only a handful of individuals survive, which are unable to reproduce due to poor health, age, sparse distribution over a large range, a lack of individuals of both sexes (in sexually reproducing species), or other reasons.

Pinpointing the extinction (or pseudoextinction) of a species requires a clear definition of that species. If it is to be declared extinct, the species in question must be uniquely identifiable from any ancestor or daughter species, or from other closely related species. Extinction of a species (or replacement by a daughter species) plays a key role in the punctuated equilibrium hypothesis of Stephen Jay Gould and Niles Eldredge.

In ecology, *extinction* is often used informally to refer to local extinction, in which a species ceases to exist in the chosen area of study, but still exists elsewhere. This phenomenon is also known as extirpation. Local extinctions may be followed by a replacement of the species taken from other locations; wolf reintroduction is an example of this. Species which are not extinct are termed extant. Those that are extant but threatened by extinction are referred to as threatened or endangered species.

An important aspect of extinction at the present time are human attempts to preserve critically endangered species, which is reflected by the creation of the conservation status "Extinct in the Wild" (EW). Species listed under this status by the International Union for Conservation of Nature (IUCN) are not known to have any living specimens in the wild, and are maintained only in zoos or other artificial environments. Some of these species are functionally extinct, as they are no longer part of their natural habitat and it is unlikely the species will ever be restored to the wild. When possible, modern zoological institutions attempt to maintain a viable population for species preservation and possible future reintroduction to the wild through use of carefully planned breeding programs.

The extinction of one species' wild population can have knock-on effects, causing further extinctions. These are also called "chains of extinction". This is especially common with extinction of keystone species.

Pseudoextinction

Descendants may or may not exist for extinct species. Daughter species that evolve from a parent species carry on most of the parent species' genetic information, and even though the parent species may become extinct, the daughter species lives on. In other cases, species have produced no new variants, or none that are able to survive the parent species' extinction. Extinction of a parent species where daughter species or subspecies are still alive is also called *pseudoextinction.*

Pseudoextinction is difficult to demonstrate unless one has a strong chain of evidence linking a living species to members of a pre-existing species. For example, it is sometimes claimed that the extinct *Hyracotherium*, which was an early horse that shares a common ancestor with the modern horse, is pseudoextinct, rather than extinct, because there are several extant species of *Equus*, including zebra and donkeys. However, as fossil species typically leave no genetic material behind, it is not possible to say whether *Hyracotherium* actually evolved into more modern horse species or simply evolved from a common ancestor with modern horses. Pseudoextinction is much easier to demonstrate for larger taxonomic groups.

Causes

As long as species have been evolving, species have been going extinct. It is estimated that over 99.9% of all species that ever lived are extinct. The average life-span of most species is 10 million years, although this varies widely between taxa. There are a variety of causes that can contribute directly or indirectly to the extinction of a species or group of species. "Just as each species is unique," write Beverly and Stephen C. Stearns, "so is each extinction ... the causes for each are varied—some subtle and complex, others obvious and simple". Most simply, any species that is unable to survive or reproduce in its environment, and unable to move to a new environment where it can do so, dies out and becomes extinct. Extinction of a species may come suddenly when an otherwise healthy species is wiped out completely, as when toxic pollution renders its entire habitat unliveable; or may occur gradually over thousands or millions of years, such as when a species gradually loses out in competition for food to better adapted competitors. Extinction may take place a long time after the events that set it in motion, a phenomenon known as extinction debt.

Assessing the relative importance of genetic factors compared to environmental ones as the causes of extinction has been compared to the nature-nurture debate. The question of whether more extinctions in the fossil record have been caused by evolution or by catastrophe is a subject of discussion; Mark Newman, the author of *Modelling Extinction* argues for a mathematical model that falls between the two positions. By contrast, conservation biology uses the extinction vortex model to classify extinctions by cause. When concerns about human extinction have been raised, for example in Sir Martin Rees' 2003 book *Our Final Hour*, those concerns lie with the effects of climate change or technological disaster.

Currently, environmental groups and some governments are concerned with the extinction of species caused by humanity, and are attempting to combat further extinctions through a variety of conservation programs. Humans can cause extinction of a species through overharvesting, pollution, habitat destruction, introduction of new predators and food competitors, overhunting, and other influences. Explosive, unsustainable human population growth is an essential cause of the extinction crisis. According to the International Union for Conservation of Nature (IUCN), 784 extinctions have been recorded since the year 1500 (to the year 2004), the arbitrary date selected to define "modern" extinctions, with many more likely to have gone unnoticed (several species have also been listed as extinct since the 2004 date).

Genetics and Demographic Phenomena

Population genetics and demographic phenomena affect the evolution, and therefore the risk of extinction, of species. Limited geographic range is the most important determinant of genus extinction at background rates but becomes increasingly irrelevant as mass extinction arises.

Natural selection acts to propagate beneficial genetic traits and eliminate weaknesses. It is nevertheless possible for a deleterious mutation to be spread throughout a population through the effect of genetic drift.

Because traits are selected and not genes, the relationship between genetic diversity and extinction risk can be complex with factors such as balancing selection, cryptic genetic variation, phenotypic plasticity, and degeneracy all playing potential roles.

A diverse or deep gene pool gives a population a higher chance of surviving an adverse change in conditions. Effects that cause or reward a loss in genetic diversity can increase the chances of extinction of a species. Population bottlenecks can dramatically reduce genetic diversity by severely limiting the number of reproducing individuals and make inbreeding more frequent. The founder effect can cause rapid, individual-based speciation and is the most dramatic example of a population bottleneck.

Genetic Pollution

Purebred wild species evolved to a specific ecology can be threatened with extinction through the process of genetic pollution—i.e., uncontrolled hybridization, introgression genetic swamping which leads to homogenization or out-competition from the introduced (or hybrid) species. Endemic populations can face such extinctions when new populations are imported or selectively bred by people, or when habitat modification brings previously isolated species into contact. Extinction is likeliest for rare species coming into contact with more abundant ones; interbreeding can swamp the rarer gene pool and create hybrids, depleting the purebred gene pool (for example, the endangered Wild water buffalo is most threatened with extinction by genetic pollution from the abundant domestic water buffalo). Such extinctions are not always apparent from morphological (non-genetic) observations. Some degree of gene flow is a normal evolutionarily process, nevertheless, hybridization (with or without introgression) threatens rare species' existence.

The gene pool of a species or a population is the variety of genetic information in its living members. A large gene pool (extensive genetic diversity) is associated with robust populations that can survive bouts of intense selection. Meanwhile, low genetic diversity reduces the range of adaptations possible. Replacing native with alien genes narrows genetic diversity within the original population, thereby increasing the chance of extinction.

Habitat Degradation

Habitat degradation is currently the main anthropogenic cause of species extinctions. The main cause of habitat degradation worldwide is agriculture, with urban sprawl, logging, mining and some fishing practices close behind. The degradation of a species' habitat may alter the fitness landscape to such an extent that the species is no longer able to survive and becomes extinct. This may occur by direct effects, such as the environment becoming toxic, or indirectly, by limiting a species' ability to compete effectively for diminished resources or against new competitor species.

Habitat degradation through toxicity can kill off a species very rapidly, by killing all living members through contamination or sterilizing them. It can also occur over longer periods at lower toxicity levels by affecting life span, reproductive capacity, or competitiveness.

Habitat degradation can also take the form of a physical destruction of niche habitats. The widespread destruction of tropical rainforests and replacement with open pastureland is widely cited as an example of this; elimination of the dense forest eliminated the infrastructure needed by many species to survive. For example, a fern that depends on dense shade for protection from direct sunlight can no longer survive without forest to shelter it. Another example is the destruction of ocean floors by bottom trawling.

Diminished resources or introduction of new competitor species also often accompany habitat degradation. Global warming has allowed some species to expand their range, bringing unwelcome competition to other species that previously occupied that area. Sometimes these new competitors are predators and directly affect prey species, while at other times they may merely outcompete vulnerable species for limited resources. Vital resources including water and food can also be limited during habitat degradation, leading to extinction.

Predation, Competition, and Disease

Before the evolution of hominids, life forms competed with each other and drove one another extinct. Recently in geologic time, Humans have been transporting animals and plants from one part of the world to another for thousands of years, sometimes deliberately (e.g., livestock released by sailors onto islands as a source of food) and sometimes accidentally (e.g., rats escaping from boats). In most cases, such introductions are unsuccessful, but when they do become established as an invasive alien species, the consequences can be catastrophic. Invasive alien species can affect native species directly by eating them, competing with them, and introducing pathogens or parasites that sicken or kill them or, indirectly, by destroying or degrading their habitat. Human populations may themselves act as invasive predators. According to the "overkill hypothesis", the swift extinction of the megafauna in areas such as Australia (40,000 years before present), North and South America (12,000 years before present), Madagascar, Hawaii (300-1000 CE), and New Zealand (1300-1500 CE), resulted from the sudden introduction of human beings to environments full of animals that had never seen them before, and were therefore completely unadapted to their predation techniques.

Coextinction

Coextinction refers to the loss of a species due to the extinction of another; for example, the extinction of parasitic insects following the loss of their hosts. Coextinction can also occur when a species loses its pollinator, or to predators in a food chain who lose their prey. "Species coextinction is a manifestation of the interconnectedness of organisms in complex ecosystems ... While coextinction may not be the most important cause of species extinctions, it is certainly an insidious one". Coextinction is especially common when a keystone species goes extinct. Models suggest that coextinction is the most common form of biodiversity loss. Coextinction allows for cascading effects across the trophic levels. Such effects are most severe in mutualistic and parasitic relationships. An example of coextinction would be the Haast's Eagle and the Moa. The Haast's Eagle is an example of a predator that became extinct because its food source became extinct. The moa were several species of flightless birds that served as a food source to the Haast's Eagle.

Climate Change

Extinction as a result of climate change has been confirmed by fossil studies. Particularly, the extinction of amphibians during the Carboniferous Rainforest Collapse, 350 million years ago. A 2003 review across 14 biodiversity research centres predicted that, because of climate change, 15–37% of land species would be "committed to extinction" by 2050. The ecologically rich areas that would potentially suffer the heaviest losses include the Cape Floristic Region, and the Caribbean Basin. These areas might see a doubling of present carbon dioxide levels and rising temperatures that could eliminate 56,000 plant and 3,700 animal species.

Mass Extinctions

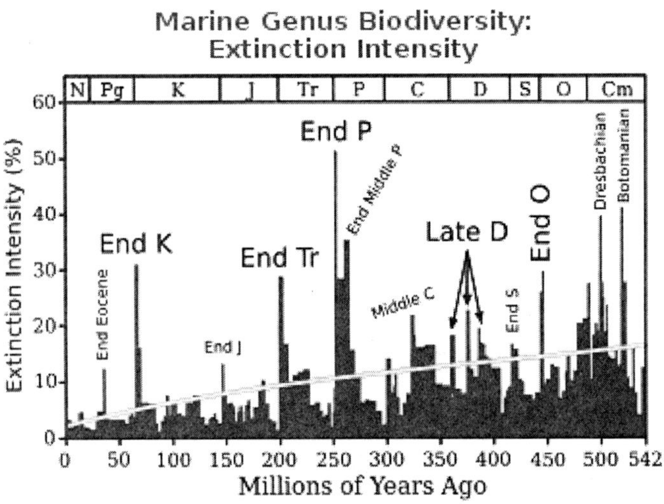

Figure : Apparent fraction of genera going extinct at any given time, as reconstructed from the fossil record.

There have been at least five mass extinctions in the history of life on earth, and four in the last 3.5 billion years in which many species have disappeared in a relatively short period of geological time. The massive eruptive event is considered to be one likely cause of the "Great Dying" about 250 million years ago, which is estimated to have killed 90% of species existing at the time. There is also evidence to suggest this event was preceded by another mass extinction known as Olson's Extinction. The Cretaceous–Tertiary extinction event occurred 65 million years ago at the end of the Cretaceous period and is best known for having wiped out non-avian dinosaurs, among many other species.

Modern Extinctions

According to a 1998 survey of 400 biologists conducted by New York's American Museum of Natural History, nearly 70 percent believed that they were currently in the early stages of a human-caused extinction, known as the Holocene extinction. In that survey, the same proportion of respondents agreed with the prediction that up to 20 percent of all living populations could become extinct within 30 years (by 2028). Biologist E. O. Wilson estimated in 2002 that if current rates of human destruction of the biosphere continue, one-half of all species of life on earth will be extinct in 100 years. More significantly the rate of species extinctions at present is estimated at 100 to 1000 times "background" or average extinction rates in the evolutionary time scale of planet Earth.

History of Scientific Understanding

In the 1750s when extinction was first described, the idea of extinction was threatening to those who held a belief in the Great Chain of Being, a theological position that did not allow for "missing links".

The possibility of extinction was not widely accepted before the 1800s. The devoted naturalist Carl Linnaeus, could "hardly entertain" the idea that humans could cause the extinction of a species. When parts of the world had not been thoroughly examined and charted, scientists could not rule out that animals found only in the fossil record were not simply "hiding" in unexplored regions of the Earth. Georges Cuvier is credited with establishing extinction as a fact in a 1796 lecture to the French Institute. Cuvier's observations of fossil bones convinced him that they did not originate in extant animals. This discovery was critical for the spread of uniformitarianism, and led to the first book publicizing the idea of evolution though Cuvier himself strongly opposed the theories of evolution advanced by Lamarck and others.

Human Attitudes and Interests

Extinction is an important research topic in the field of zoology, and biology in general, and has also become an area of concern outside the scientific community. A number of organisations, such as the Worldwide Fund for Nature, have been created with the goal of preserving species from extinction. Governments have attempted, through enacting laws, to avoid habitat destruction, agricultural over-harvesting, and pollution. While many human-caused extinctions have been accidental, humans have also engaged in the deliberate destruction of some species, such as dangerous viruses, and the total destruction of other problematic species has been suggested. Other species were deliberately driven to extinction, or nearly so, due to poaching or because they were "undesirable", or to push for other human agendas. One example was the near extinction of the American bison, which was nearly wiped out by mass hunts sanctioned by the United States government, in order to force the removal of Native Americans, many of whom relied on the bison for food.

Biologist Bruce Walsh of the University of Arizona states three reasons for scientific interest in the preservation of species; genetic resources, ecosystem stability, and ethics; and today the scientific community "stress[es] the importance" of maintaining biodiversity.

In modern times, commercial and industrial interests often have to contend with the effects of production on plant and animal life. However, some technologies with minimal, or no, proven harmful effects on *Homo sapiens* can be devastating to wildlife (for example, DDT). Biogeographer Jared Diamond notes that while big business may label environmental concerns as "exaggerated", and often cause "devastating damage", some corporations find it in their interest to adopt good conservation practices, and even engage in preservation efforts that surpass those taken by national parks.

Governments sometimes see the loss of native species as a loss to ecotourism, and can enact laws with severe punishment against the trade in native species in an effort to prevent extinction in the wild. Nature preserves are created by governments as a means to provide continuing habitats to species crowded by human expansion. The 1992 Convention on Biological Diversity has resulted in international Biodiversity Action Plan programmes, which attempt to provide comprehensive guidelines for government biodiversity conservation. Advocacy groups, such as The Wildlands Project and the Alliance for Zero Extinctions, work to educate the public and pressure governments into action.

People who live close to nature can be dependent on the survival of all the species in their environment, leaving them highly exposed to extinction risks. However, people prioritize day-to-day survival over species conservation; with human overpopulation in tropical developing countries, there has been enormous pressure on forests due to subsistence agriculture, including slash-and-burn agricultural techniques that can reduce endangered species's habitats.

Michael Levin argues, "The very fact that a species is near extinction implies that its final demise will have negligible impact."

Planned Extinction

Humans have aggressively worked toward the extinction of many species of viruses and bacteria in the cause of disease eradication.

Implemented:

- The smallpox virus is now extinct in the wild—although samples are retained in laboratory settings.
- The polio virus is now confined to small parts of the world as a result of human efforts to prevent the disease it causes.
- Guinea worm
- Rinderpest.

Proposed

Olivia Judson is one of six modern scientists to have advocated the deliberate extinction of specific species. Her September 25, 2003 *New York Times* article, "A Bug's Death", advocates "specicide" of thirty mosquito species through the introduction of a genetic element, capable of inserting itself into another crucial gene, to create recessive "knockout genes".

Her arguments for doing so are that the *Anopheles* mosquitoes (which spread malaria) and *Aedes* mosquitoes (which spread dengue fever, yellow fever, elephantiasis, and other diseases) represent only 30 species; eradicating these would save at least one million human lives per annum at a cost of reducing the genetic diversity of the family Culicidae by only 1%. She further argues that since species become extinct "all the time" the disappearance of a few more will not destroy the ecosystem: "We're not left with a wasteland every time a species vanishes. Removing one species sometimes causes shifts in the populations of other species — but different need not mean worse." In addition, anti-malarial and mosquito control programs offer little realistic hope to the 300 million people in developing nations who will be infected with acute illnesses this year. Although trials are ongoing, she writes that if they fail: "We should consider the ultimate swatting."

Cloning

Ongoing technological advances have encouraged the hypothesis that by using DNA from the remains of an extinct species, through the process of cloning, the species may be "brought back to life". Proposed targets for cloning include the dinosaurs, the mammoth, thylacine, and the Pyrenean Ibex. In order for such a program to succeed, a sufficient number of individuals would have to be cloned, from the DNA of different individuals (in the case of sexually reproducing organisms) to create a viable population. Though bioethical and philosophical objections have been raised, the cloning of extinct creatures seems a viable outcome of the continuing advancements in our science and technology.

In 2003, scientists attempted to clone the extinct Pyrenean Ibex (*C. p. pyrenaica*). This initial attempt failed; of the 285 embryos reconstructed, 54 were transferred to 12 mountain goats and mountain goat-domestic goat hybrids, but only two survived the initial two months of gestation before they too died. In 2009, a second attempt was made to clone the Pyrenean Ibex; one clone was born alive, but died seven minutes later, due to physical defects in the lungs.

The concept of cloning extinct species was thought to be first popularized by the successful novel and subsequent film *Jurassic Park*, though it may have been first used in John Brosnan's 1984 novel *Carnosaur*, then in F. Paul Wilson's 1989 novel *Dydeetown World,* and later in Piers Anthony's 1990 novel *Balook*, which featured the resurrection of a *Baluchitherium.*

- Endling
- Extinction risk from global warming
- Gene pool
- Genetic erosion
- Genetic pollution
- Genocide
- Habitat fragmentation
- IUCN Red List

- IUCN Red List of extinct species for a list by taxonomy
- Ecological importance of bees
- List of extinct animals
- List of extinct plants
- Living Planet Index
- Mass extinction
- Overexploitation
- Red List Index
- Refugium (population biology)
- Speciation
- Timeline of extinctions
- Voluntary Human Extinction Movement
- Dinosaurs.

Chapter 4

Sustainability Standards and Certification

The term "sustainability standards" refers to a voluntary, usually third party-assessed, norms and standards relating to environmental, social, ethical and food safety issues, adopted by companies to demonstrate the performance of their organisations or products in specific areas. There are perhaps up to 500 such standards and the pace of introduction has increased in the last decade.

The trend started in the late 1980s and 90s with the introduction of Ecolabels and standards for Organic food and other products. In recent years, numerous standards have been established and adopted in the food industry in particular. Most of them refer to the triple bottom line of environmental quality, social equity, and economic prosperity. The basic premise of sustainability standards is two fold. Firstly, they emerged in areas where national and global legislation was weak but where the consumer and NGO movements around the globe demanded action. For example, campaigns by Global Exchange and other NGOs against the purchase of goods from "sweatshop" factories by the likes of Nike, Inc., Levi Strauss & Co. and other leading brands led to the emergence of social welfare standards like the SA8000 and others. Secondly, leading brands selling to both consumers and to the B2B supply chain may wish to demonstrate the environmental or organic merits of their products, which has led to the emergence of hundreds of ecolabels, organic and other standards.

A leading example of a consumer standard is the Fairtrade movement, administered by FLO International and exhibiting huge sales growth around the world for ethically sourced produce. An example of a B2B standard which has grown tremendously in the last few years is the Forest Stewardship Council's standard (FSC) for forest products made from sustainably harvested trees. However, the line between consumer and B2B sustainability standards is becoming blurred, with leading trade buyers increasingly demanding Fairtrade certification, for example, and consumers increasingly recognising the FSC mark. A list of 350 or so of the main global sustainability standards may be found here.

Different Sustainability Standards

Numerous sustainability standards have been developed in recent years to address issues of environmental quality, social equity, and economic prosperity of global production and trade practices. Despite similarities in major goals and certification procedures, there are some significant differences in terms of their historical development, target groups of adopters, geographical diffusion, and emphasis on environmental, social or economic issues.

Fairtrade

The Fairtrade label was developed in the late 1980s by a Dutch development agency in collaboration with Mexican farmers. The initiative performs development work and promotes its political vision of an alternative economy, seeing its main objective in empowering small producers and providing these with access to and improving their position on global markets.

The most distinguishing feature of the Fairtrade label is the guarantee of a minimum price and a social premium that goes to the cooperative and not to the producers directly. Recently, Fairtrade also adopted environmental objectives as part of their certification system. The most prominent standard and associated certification system within the Fairtrade movement is administered by FLO International, a Bonn based non profit organisation which controls the Fairtrade Certification Mark, a standard to which over 800 producer organisations have been certified to as of 2011

Rainforest Alliance

The Rainforest Alliance was created in the late 1980s from a social movement and is committed to conserving rainforests and their biodiversity. One key element of the standard is the compulsory elaboration and implementation of a detailed plan for the development of a sustainable farm management system so as to assist wildlife conservation. Another objective is to improve workers' welfare by establishing and securing sustainable livelihoods. Producer prices may carry a premium. Yet instead of guaranteeing a fixed floor price, the standard seeks to improve the economic situation of producers through higher yields and enhanced cost efficiency.

Utz Certified

Utz Certified (formerly Utz Kapeh) was co-founded by the Dutch coffee roaster Ahold Coffee Company in 1997. It aims to create an open and transparent marketplace for socially and environmentally responsible agricultural products. Instruments include the UTZ Traceability System and the UTZ Code of Conduct. The traceability system makes certified products traceable from producer to final buyer and has stringent chains of custody requirements. The UTZ Code of Conduct emphasizes both environmental practices (e.g. biodiversity conservation, waste handling and water use) and social benefits (e.g. access to medical care, access to sanitary facilities at work).

Organic

The Organic standard was developed in the 1970s and is based on IFOAM Basic Standards. IFOAM stands for International Federation of Organic Agriculture Movements and is the leading global umbrella organisation for the organic farming movement. The IFOAM Basic Standards provide a framework of minimum requirements, including the omission of agrochemicals such as pesticides and chemical-synthetic fertilizers. The use of animal feeds is also strictly regulated. Genetic engineering and the use of genetically modified organisms (GMOs) are forbidden.

Origin of Global Standards

Most sustainability standards that are being adopted today were initiated by social movements in particular countries, such as Rainforest Alliance in the United States and Fairtrade in the Netherlands. Other standards were initiated by individual companies, such as Utz Certified (Ahold), Starbucks C.A.F.E. (Starbucks), and Nespresso AAA (Nespresso). Some standards were launched by coalitions of private firms, development agencies, NGOs, and other stakeholders. For example, the Common Code for the Coffee Community (4C) was initiated by an alliance of main coffee roasters, including Kraft Foods, Sara Lee and Nestle, assisted by the German Agency for Technical Cooperation and Development GIZ.

There are important facilitating conditions for standards development. One important factor are institutional and economic conditions in particular consuming countries, e.g. awareness of environmental and social issues among consumers as well as institutional support for sustainability initiatives. One example for the former are long-term sustainability movements in the Netherlands;

one example for the latter are government-funded development programs around sustainability in Germany. Another important factor are supporting political and economic conditions in producing countries, as well as successful pilot projects involving multiple stakeholders, such as development agencies, NGOs, local producers, and multinational corporations.

For example, the Fairtrade standard was developed based on pilot projects with Mexican farmers. 4C builds on development projects in Peru, Colombia and Vietnam, involving GIZ, major coffee roasters, and local producers.

Sustainable Development

Figure : Solar power towers utilise the natural resource of the Sun, and are a renewable energy source. From left: PS10 and PS20 solar towers.

Sustainable development (SD) is a pattern of resource use, that aims to meet human needs while preserving the environment so that these needs can be met not only in the present, but also for generations to come (sometimes taught as ELF-Environment, Local people, Future). The term was used by the Brundtland Commission which coined what has become the most often-quoted definition of sustainable development as development that "meets the needs of the present without compromising the ability of future generations to meet their own needs." Sustainable development ties together concern for the carrying capacity of natural systems with the social challenges facing humanity. As early as the 1970s "sustainability" was employed to describe an economy "in equilibrium with basic ecological support systems." Ecologists have pointed to *The Limits to Growth*, and presented the alternative of a "steady state economy" in order to address environmental concerns.

The field of sustainable development can be conceptually broken into three constituent parts: environmental sustainability, economic sustainability and sociopolitical sustainability.

Definition of Sustainable Development

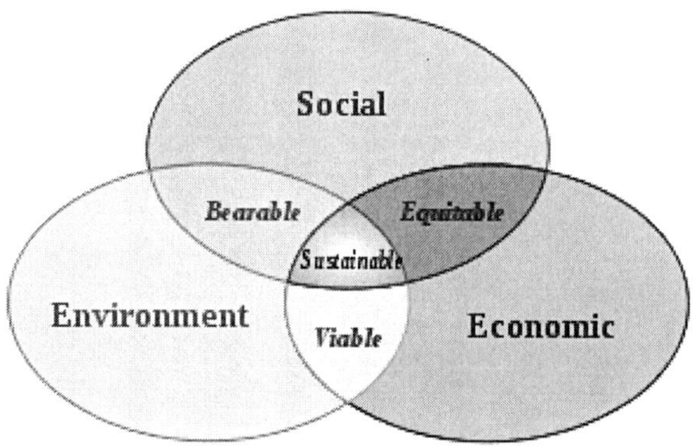

Figure : Scheme of sustainable development: at the confluence of three constituent parts.(2006)

In 1987, the United Nations released the Brundtland Report, which included what is now one of the most widely recognised definitions:

"Sustainable development is development that meets the needs of the present without compromising the ability of future generations to meet their own needs." It contains within it two key concepts:

- the concept of 'needs', in particular the essential needs of the world's poor, to which overriding priority should be given; and
- the idea of limitations imposed by the state of technology and social organisation on the environment's ability to meet present and future needs."

The United Nations 2005 World Summit Outcome Document refers to the "interdependent and mutually reinforcing pillars" of sustainable development as economic development, social development, and environmental protection. Based on the triple bottom line, numerous sustainability standards and certification systems have been established in recent years, in particular in the food industry. Well-known standards include Organic, Rainforest Alliance, Fairtrade, Utz Certified, Bird Friendly, and The Common Code for the Coffee Community.

Indigenous peoples have argued, through various international forums such as the United Nations Permanent Forum on Indigenous Issues and the Convention on Biological Diversity, that there are *four* pillars of sustainable development, the fourth being cultural. *The Universal Declaration on Cultural Diversity* (UNESCO, 2001) further elaborates the concept by stating that "...cultural diversity is as necessary for humankind as biodiversity is for nature"; it becomes "one of the roots of development understood not simply in terms of economic growth, but also as a means to achieve a more satisfactory intellectual, emotional, moral and spiritual existence". In this vision, cultural diversity is the fourth policy area of sustainable development.

A useful articulation of the values and principles of sustainability can be found in the Earth Charter. It offers an integrated vision and definition of strong sustainability. The document, an ethical framework for a sustainable world, was developed over several years after the Rio Earth Summit in 1992 and launched officially in 2000. The Charter derives its legitimacy from the participatory process in which it was drafted, which included contributions from hundreds of organisations and thousands of individuals, and from its use since 2000 by thousands of organisations and individuals that have been using the Earth Charter as an educational instrument and a policy tool.

Economic Sustainability: Agenda 21 clearly identified information, integration, and participation as key building blocks to help countries achieve development that recognises these interdependent pillars. It emphasises that in sustainable development everyone is a user and provider of information. It stresses the need to change from old sector-centred ways of doing business to new approaches that involve cross-sectoral co-ordination and the integration of environmental and social concerns into all development processes. Furthermore, Agenda 21 emphasises that broad public participation in decision making is a fundamental prerequisite for achieving sustainable development.

According to Hasna Vancock, sustainability is a process which tells of a development of all aspects of human life affecting sustenance. It means resolving the conflict between the various competing goals, and involves the simultaneous pursuit of economic prosperity, environmental quality and social equity famously known as three dimensions (triple bottom line) with the resultant vector being technology, hence it is a continually evolving process; the 'journey' (the process of achieving sustainability) is of course vitally important, but only as a means of getting to the destination (the desired future state).

However, the 'destination' of sustainability is not a fixed place in the normal sense that we understand destination. Instead, it is a set of wishful characteristics of a future system. The concept has included notions of weak sustainability, strong sustainability and deep ecology.

Green development is generally differentiated from sustainable development in that Green development prioritizes what its proponents consider to be environmental sustainability over economic and cultural considerations. Proponents of Sustainable Development argue that it provides a context in which to improve overall sustainability where cutting edge Green development is unattainable. For example, a cutting edge treatment plant with extremely high maintenance costs may not be sustainable in regions of the world with fewer financial resources. An environmentally ideal plant that is shut down due to bankruptcy is obviously less sustainable than one that is maintainable by the community, even if it is somewhat less effective from an environmental standpoint.

Some research activities start from this definition to argue that the environment is a combination of nature and culture. The Network of Excellence "Sustainable Development in a Diverse World", sponsored by the European Union, integrates multidisciplinary capacities and interprets cultural diversity as a key element of a new strategy for sustainable development.

In fact, some researchers and institutions have even pointed out that a fourth dimension should be added to the three dimensions of sustainable development, since these three dimensions do not seem to be enough to reflect the complexity of contemporary society. In this context, the Agenda 21 for culture and the United Cities and Local Governments (UCLG) Executive Bureau lead the preparation of the policy statement "Culture: Fourth Pillar of Sustainable Development", passed on 17 November 2010, in the framework of the World Summit of Local and Regional Leaders – 3rd World Congress of UCLG, held in Mexico City.

This document inaugurates a new perspective and points to the relation between culture and sustainable development through a dual approach: developing a solid cultural policy and advocating a cultural dimension in all public policies.

Still other researchers view environmental and social challenges as opportunities for development action. This is particularly true in the concept of sustainable enterprise that frames these global needs as opportunities for private enterprise to provide innovative and entrepreneurial solutions. This view is now being taught at many business schools including the Centre for Sustainable Global Enterprise at Cornell University and the Erb Institute for Global Sustainable Enterprise at the University of Michigan.

The United Nations Division for Sustainable Development lists the following areas as coming within the scope of sustainable development:

Sustainable development is an eclectic concept, as a wide array of views fall under its umbrella. The concept has included notions of weak sustainability, strong sustainability and deep ecology. Different conceptions also reveal a strong tension between ecocentrism and anthropocentrism. Many definitions and images (Visualizing Sustainability) of sustainable development coexist. Broadly defined, the sustainable development mantra enjoins current generations to take a systems approach to growth and development and to manage natural, produced, and social capital for the welfare of their own and future generations.

During the last ten years, different organisations have tried to measure and monitor the proximity to what they consider sustainability by implementing what has been called sustainability metrics and indices.

Sustainable development is said to set limits on the developing world. While current first world countries polluted significantly during their development, the same countries encourage third world countries to reduce pollution, which sometimes impedes growth. Some consider that the implementation of sustainable development would mean a reversion to pre-modern lifestyles.

Others have criticized the overuse of the term:

> *"[The] word sustainable has been used in too many situations today, and ecological sustainability is one of those terms that confuse a lot of people. You hear about sustainable development, sustainable growth, sustainable economies, sustainable societies, sustainable agriculture. Everything is sustainable (Temple, 1992)."*

Environmental Sustainability

Figure : Water is an important natural resource that covers 71% of the Earth's surface. Image is the Earth photographed from Apollo 17.

Environmental sustainability is the process of making sure current processes of interaction with the environment are pursued with the idea of keeping the environment as pristine as naturally possible based on ideal-seeking behaviour.

An "unsustainable situation" occurs when natural capital (the sum total of nature's resources) is used up faster than it can be replenished. Sustainability requires that human activity only uses nature's resources at a rate at which they can be replenished naturally. Inherently the concept of sustainable development is intertwined with the concept of carrying capacity. Theoretically, the long-term result of environmental degradation is the inability to sustain human life. Such degradation on a global scale could imply extinction for humanity.

Economic Sustainability

The Venn diagram of sustainable development shown above has many versions, but was first used by economist Edward Barbier (1987). However, Pearce, Barbier and Markandya (1989) criticized the Venn approach due to the intractability of operationalizing separate indices of economic, environmental, and social sustainability and somehow combining them. They also noted that the Venn approach was inconsistent with the Brundtland Commission Report, which emphasized the interlinkages between economic development, environmental degradation, and population pressure instead of three objectives. Economists have since focused on viewing the economy and the environment as a single interlinked system with a unified valuation methodology (Hamilton 1999, Dasgupta 2007). Intergenerational equity can be incorporated into this approach, as has become common in economic valuations of climate change economics (Heal,2009). Ruling out discrimination against future generations and allowing for the possibility of renewable alternatives to petro-chemicals and other non-renewable resources, efficient policies are compatible with increasing human welfare, eventually reaching a golden-rule steady state (Ayong le Kama, 2001 and Endress et al.2005). Thus the three pillars of sustainable development are interlinkages, intergenerational equity, and dynamic efficiency (Stavins, et al. 2003).

Arrow et al. (2004) and other economists (e.g. Asheim,1999 and Pezzey, 1989 and 1997) have advocated a form of the weak criterion for sustainable development – the requirement than the wealth of a society, including human-capital, knowledge-capital and natural-capital (as well as produced capital) not decline over time. Others, including Barbier 2007, continue to contend that strong sustainability – non-depletion of essential forms of natural capital – may be appropriate.

Three Types of Capital in Sustainable Development

The sustainable development debate is based on the assumption that societies need to manage three types of capital (economic, social, and natural), which may be non-substitutable and whose consumption might be irreversible. Daly (1991), for example, points to the fact that natural capital can not necessarily be substituted by economic capital. While it is possible that we can find ways to replace some natural resources, it is much more unlikely that they will ever be able to replace eco-system services, such as the protection provided by the ozone layer, or the climate stabilizing function of the Amazonian forest. In fact natural capital, social capital and economic capital are often complementarities.

A further obstacle to substitutability lies also in the multi-functionality of many natural resources. Forests, for example, not only provide the raw material for paper (which can be substituted quite easily), but they also maintain biodiversity, regulate water flow, and absorb CO2.

Another problem of natural and social capital deterioration lies in their partial irreversibility. The loss in biodiversity, for example, is often definite. The same can be true for cultural diversity. For example with globalisation advancing quickly the number of indigenous languages is dropping at alarming rates. Moreover, the depletion of natural and social capital may have non-linear consequences.

Consumption of natural and social capital may have no observable impact until a certain threshold is reached. A lake can, for example, absorb nutrients for a long time while actually increasing its productivity. However, once a certain level of algae is reached lack of oxygen causes the lake's ecosystem to break down suddenly.

Market Failure

If the degradation of natural and social capital has such important consequence the question arises why action is not taken more systematically to alleviate it. Cohen and Winn (2007) point to four types of market failure as possible explanations: First, while the benefits of natural or social capital depletion can usually be privatized the costs are often externalized (i.e. they are borne not by the party responsible but by society in general). Second, natural capital is often undervalued by society since we are not fully aware of the real cost of the depletion of natural capital. Information asymmetry is a third reason—often the link between cause and effect is obscured, making it difficult for actors to make informed choices. Cohen and Winn close with the realisation that contrary to economic theory many firms are not perfect optimizers. They postulate that firms often do not optimize resource allocation because they are caught in a "business as usual" mentality.

The Business Case for Sustainable Development

The most broadly accepted criterion for corporate sustainability constitutes a firm's efficient use of natural capital. This eco-efficiency is usually calculated as the economic value added by a firm in relation to its aggregated ecological impact. This idea has been popularised by the World Business Council for Sustainable Development (WBCSD) under the following definition: "Eco-efficiency is achieved by the delivery of competitively priced goods and services that satisfy human needs and bring quality of life, while progressively reducing ecological impacts and resource intensity throughout the life-cycle to a level at least in line with the earth's carrying capacity." (DeSimone and Popoff, 1997)

Similar to the eco-efficiency concept but so far less explored is the second criterion for corporate sustainability. Socio-efficiency describes the relation between a firm's value added and its social impact. Whereas, it can be assumed that most corporate impacts on the environment are negative (apart from rare exceptions such as the planting of trees) this is not true for social impacts. These can be either positive (e.g. corporate giving, creation of employment) or negative (e.g. work accidents, mobbing of employees, human rights abuses). Depending on the type of impact socio-efficiency thus either tries to minimise negative social impacts (i.e. accidents per value added) or maximise positive social impacts (i.e. donations per value added) in relation to the value added.

Both eco-efficiency and socio-efficiency are concerned primarily with increasing economic sustainability. In this process they instrumentalize both natural and social capital aiming to benefit from win-win situations. However, as Dyllick and Hockerts point out the business case alone will not be sufficient to realise sustainable development. They point towards eco-effectiveness, socio-effectiveness, sufficiency, and eco-equity as four criteria that need to be met if sustainable development is to be reached..

Sustainable Agriculture

Sustainable Agriculture can be simply defined as environmentally friendly methods of farming that allow the production of crops or livestock without damage to the farm as an ecosystem. Apart from this, it also prevents the adverse effects on soil, water supplies, biodiversity, or other surrounding natural resources. The concept of sustainable agriculture is an intergenerational one in which we pass on a conserved or improved natural resource base instead of one which has been depleted or polluted.

The Elements of Sustainable Agriculture

Agroforestry : According to the World Agroforestry Centre, Agroforestry is a collective name for land use systems and practices in which woody perennials are deliberately integrated with crops and/ or animals on the same land management unit. The integration can be either in a spatial mixture or in a temporal sequence. There are normally both ecological and economic interactions between woody and non-woody components in agroforestry.

Mixed Farming: Many farmers in tropical & temperate countries survive by managing a mix of different crops or animals. The best known form of mixing occurs probably where crop residues are used to feed the animals and the excreta from animals are used as nutrients for the crop. Other forms of mixing takes place where graving under fruit trees keeps the grass short or where manure from pigs is used to feed the fish. Mixed farming exists in many forms depending on external and internal factors. Mixed Farming provides farmers with a) an opportunity to diversify risk from single-crop production; (b) to use labour more efficiently; (c) to have a source of cash for purchasing farm inputs; (d) to add value to crop or crop by-product; (e) combining crops and livestocks.

Multiple Cropping : The process of growing two or more crops in the same piece of land, during the same season is called Multiple Cropping. It can be rightly called a form of polyculture. It can be – (a) Double Cropping (the practice where the second crop is planted after the first has been harvested); (b) Relay Cropping (the practice where a second crop is started along with the first one, before it is harvested).

Crop Rotation : The process of growing two or more dissimilar or unrelated crops in the same piece of land in different seasons is known as Crop Rotation. This process could be adopted as it comes with a series of benefits like – (a) avoid the build up of pests that often occurs when one species is continuously cropped; (b) the traditional element of crop rotation is the replenishment of nitrogen through the use of green manure in sequence with cereals and other crops; (c) Crop rotation can also improve soil structure and fertility by alternating deep-rooted and shallow-rooted plants; (d) it is a component of polyculture.

Criticisms

The concept of "Sustainable Development" raises several critiques at different levels.

Consequences

Figure : The retreat of Aletsch Glacier in the Swiss Alps (situation in 1979, 1991 and 2002) due to warming.

John Baden views the notion of sustainable development as dangerous because the consequences have unknown effects. He writes: "In economy like in ecology, the interdependence rule applies. Isolated actions are impossible. A policy which is not carefully enough thought will carry along various perverse and adverse effects for the ecology as much as for the economy. Many suggestions to save our environment and to promote a model of 'sustainable development' risk indeed leading to reverse effects." Moreover, he evokes the bounds of public action which are underlined by the public choice theory: the quest by politicians of their own interests, lobby pressure, partial disclosure etc. He develops his critique by noting the vagueness of the expression, which can cover anything .

It is a gateway to interventionist proceedings which can be against the principle of freedom and without proven efficacy. Against this notion, he is a proponent of private property to impel the producers and the consumers to save the natural resources. According to Baden, "the improvement of environment quality depends on the market economy and the existence of legitimate and protected property rights." They enable the effective practice of personal responsibility and the development of mechanisms to protect the environment. The State can in this context "create conditions which encourage the people to save the environment."

Vagueness of the Term

Figure : A sewage treatment plant that uses environmentally friendly solar energy, located at Santuari de Lluc monastery.

Some criticize the term "sustainable development", stating that the term is too vague. For example, both Jean-Marc Jancovici and the philosopher Luc Ferry express this view. The latter writes about sustainable development: "I know that this term is obligatory, but I find it also absurd, or rather so vague that it says nothing." Luc Ferry adds that the term is trivial by a proof of contradiction: "who would like to be a proponent of an "untenable development! Of course no one! [..] The term is more charming than meaningful. [..] Everything must be done so that it does not turn into Russian-type administrative planning with ill effects." Sustainable development has become obscured by conflicting world views, the expansionist and the ecological, and risks being co-opted by individuals and institutions that perpetuate many aspects of the expansionist model.

Basis

Sylvie Brunel, French geographer and specialist of the Third World, develops in *A qui profite le développement durable* (Who benefits from sustainable development?) (2008) a critique of the basis of sustainable development, with its binary vision of the world, can be compared to the Christian vision of Good and Evil, an idealized nature where the human being is an animal like the others or even an alien. Nature – as Rousseau thought – is better than the human being. It is a parasite, harmful for the nature. But the human is the one who protects the biodiversity, where normally only the strong survive.

Moreover, she thinks that the core ideas of sustainable development are a hidden form of protectionism by developed countries impeding the development of the other countries. For Sylvie Brunel, sustainable development serves as a pretext for protectionism and "I have the feeling that sustainable development is perfectly helping out capitalism".

"De-growth"

The proponents of the de-growth reckon that the term of sustainable development is an oxymoron. According to them, on a planet where 20% of the population consumes 80% of the natural resources, a sustainable development cannot be possible for this 20%: "According to the origin of the concept of sustainable development, a development which meets the needs of the present without compromising the ability of future generations to meet their own needs, the right term for the developed countries should be a sustainable de-growth".

Measurability

In 2007 a report for the U.S. Environmental Protection Agency stated: "While much discussion and effort has gone into sustainability indicators, none of the resulting systems clearly tells us whether our society is sustainable. At best, they can tell us that we are heading in the wrong direction, or that our current activities are not sustainable. More often, they simply draw our attention to the existence of problems, doing little to tell us the origin of those problems and nothing to tell us how to solve them." Nevertheless a majority of authors assume that a set of well defined and harmonised indicators is the only way to make sustainability tangible. Those indicators are expected to be identified and adjusted through empirical observations (trial and error). The most common critiques are related to issues like data quality, comparability, objective function and the necessary resources. However a more general criticism is coming from the project management community: How can a sustainable development be achieved at global level if we cannot monitor it in any single project?

The Cuban-born researcher and entrepreneur Sonia Bueno suggests an alternative approach that is based upon the integral, long-term cost-benefit relationship as a measure and monitoring tool for the sustainability of every project, activity or enterprise. Furthermore this concept aims to be a practical guideline towards sustainable development following the principle of conservation and increment of value rather than restricting the consumption of resources.

Organisations and Research

- 2010 Biodiversity Indicators Partnership
- 2010 International Year of Biodiversity
- Afrique verte
- Agronomy for Sustainable Development
- Appropedia
- Centre for Sustainable Enterprise and Regional Competitiveness (SERC)
- Centre for Environment, Technology & Development, Malaysia (Cetdem)
- Dashboard of Sustainability
- Earth Charter
- Global Map
- Greenhouse Development Rights
- Institute for Environment and Sustainability (IES)
- Institute for Trade, Standards and Sustainable Development (ITSSD)

- International Institute for Environment and Development
- International Institute for Sustainable Development (IISD)
- International Mountain Day - December 11
- International Organisation for Sustainable Development
- National Centre for Appropriate Technology
- National Strategy for a Sustainable America
- Solar Net International
- Stakeholder Forum for a Sustainable Future
- Sustainable Tourism CRC
- The Earth Institute
- TERI – The Energy and Resources Institute
- The Sustainable Urban Development Network (SUD-Net)
- The Venus Project
- United Nations Decade of Education for Sustainable Development
- World Cities Summit.

Sustainable Living

Sustainable living is a lifestyle that attempts to reduce an individual's or society's use of the Earth's natural resources and his/her own resources. Practitioners of sustainable living often attempt to reduce their carbon footprint by altering methods of transportation, energy consumption and diet. Proponents of sustainable living aim to conduct their lives in ways that are consistent with sustainability, in natural balance and respectful of humanity's symbiotic relationship with the Earth's natural ecology and cycles. The practice and general philosophy of ecological living is highly interrelated with the overall principles of sustainable development.

Lester R. Brown, a prominent environmentalist and founder of the Worldwatch Institute and Earth Policy Institute, describes sustainable living in the 21st century as "shifting to a renewable energy-based, reuse/recycle economy with a diversified transport system."

Definition

Sustainable living is fundamentally the application of sustainability to lifestyle choice and decisions. Sustainability itself is expressed as meeting present ecological, societal, and economical needs without compromising these factors for future generations Sustainable living can therefore be described as living within the innate carrying capacities defined by these factors.

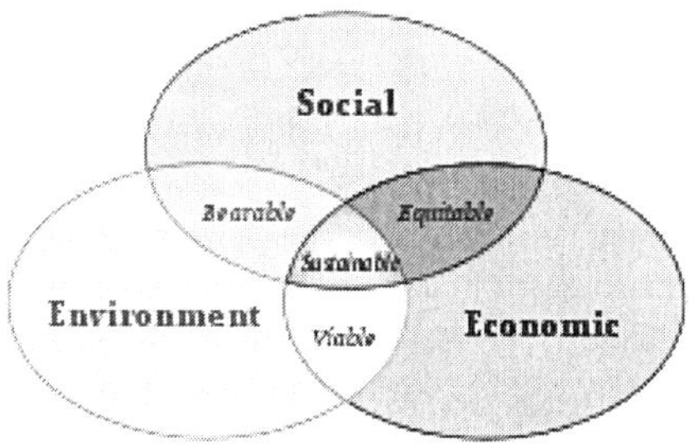

Figure : The three pillars of sustainability.

Sustainable design and sustainable development are critical factors to sustainable living. Sustainable design encompasses the development of appropriate technology, which is a staple of sustainable living practices. Sustainable development in turn is the use of these technologies in infrastructure. Sustainable architecture and agriculture are the most common examples of this practice.

History

- 1954 The publication of *Living the Good Life* by Helen and Scott Nearing marked the beginning of the modern day sustainable living movement. The publication paved the way for the "back-to-the-land movement" in the late 1960s and early 1970s.
- 1962 The publication of *Silent Spring* by Rachel Carson marked another major milestone for the sustainability movement.
- 1972 Donella Meadows wrote the international bestseller *The Limits to Growth*, which reported on a study of long-term global trends in population, economics and the environment. It sold millions of copies and was translated into 28 languages.
- 1973 E. F. Schumacher published a collection of essays on shifting towards sustainable living through the appropriate use of technology in his book *Small is Beautiful.*
- 1992–2002 The United Nations held a series of conferences, which focused on increasing sustainability within societies in order to conserve the Earth's natural resources. The Earth Summit conferences were held in 1992, 1972 and 2002.

- 2007 the United Nations published *Sustainable Consumption and Production, Promoting Climate-Friendly Household Consumption Patterns*, which promoted sustainable lifestyles in communities and homes.

Shelter

Sustainable homes are built using sustainable methods, materials, and facilitate green practices, enabling a sustainable lifestyle. Their construction and maintenance have neutral impacts on the Earth. Oftentimes, if necessary, they are close in proximity to essential services such as grocery stores, schools, daycares, work, or public transit making it possible to commit to sustainable transportation choices. Sometimes, they are off-the-grid homes that do not require any public energy, water, or sewer service.

If not off-the-grid, sustainable homes may be linked to a grid supplied by a power plant that is using sustainable power sources, buying power as is normal convention. Additionally, sustainable homes may be connected to a grid, but generate their own electricity through renewable means and sell any excess to a utility. There are two common methods to approaching this option: net metring and double metring.

Net metring uses the common metre that is installed in most homes, running forward when power is used from the grid, and running backward when power is put into the grid (which allows them to "net" out their total energy use, putting excess energy into the grid when not needed, and using energy from the grid during peak hours, when you may not be able to produce enough immediately). Power companies can quickly purchase the power that is put back into the grid, as it is being produced. Double metring involves installing two metres: one measuring electricity consumed, the other measuring electricity created. Additionally, or in replace of selling their renewable energy, sustainable home owners may choose to bank their excess energy by using it to charge batteries. This gives them the option to use the power later during less favourable power-generating times (i.e.: night-time, when there has been no wind, etc.), and to be completely independent of the electrical grid.

Sustainably designed houses are generally sited so as to create as little of a negative impact on the surrounding ecosystem as possible, oriented to the sun so that it creates the best possible microclimate (typically, the long axis of the house or building should be oriented east-west), and provide natural shading or wind barriers where and when needed, among many other considerations. The design of a sustainable shelter affords the options it has later (i.e.: using passive solar lighting and heating, creating temperature buffer zones by adding porches, deep overhangs to help create favourable microclimates, etc.)

Sustainably constructed houses involve environmentally-friendly management of waste building materials such as recycling and composting, use non-toxic and renewable, recycled, reclaimed, or low-impact production materials that have been created and treated in a sustainable fashion (such as using organic or water-based finishes), use as much locally available materials and tools as possible so as to reduce the need for transportation, and use low-impact production methods (methods that minimise effects on the environment).

Many materials can be considered a "green" material until its background is revealed. Any material that has used toxic or carcinogenic chemicals in its treatment or manufacturing (such as formaldehyde in glues used in woodworking), has travelled extensively from its source or manufacturer, or has been cultivated or harvested in an unsustainable manner might not be considered green. In order for any material to be considered green, it must be resource efficient, not compromise indoor air quality or water conservation, and be energy efficient (both in processing and when in use in the shelter). Resource efficiency can be achieved by using as much recycled content, reusable or recyclable content, materials that employ recycled or recyclable packaging, locally available material, salvaged or remanufactured material, material that employs resource efficient manufacturing, and long-lasting material as possible.

List of some sustainable materials:

- Adobe
- Bamboo
- Cellulose insulation
- Cob
- Composite wood (when made from reclaimed hardwood sawdust and reclaimed or recycled plastic)
- Cordwood
- Cork
- Hemp
- Insulating concrete forms
- Lime render

- Linoleum
- Lumber from Forest Stewardship Council approved sources
- Natural Rubber
- Natural fibre (coir, wool, jute, etc.)
- Organic cotton insulation
- Papercrete
- Rammed Earth
- Reclaimed stone
- Reclaimed brick
- Recycled metal
- Recycled concrete
- Recycled paper
- Soy-based adhesive
- Soy insulation
- Straw Bale
- Structural insulated panel.

Insulation of a sustainable home is important because of the energy it conserves throughout the life of the home. Well insulated walls and lofts using green materials are a must as it reduces or, in combination with a house that is well designed, eliminates the need for heating and cooling altogether. Installation of insulation varies according to the type of insulation being used. Typically, lofts are insulated by strips of insulating material laid between rafters. Walls with cavities are done in much the same manner. For walls that do not have cavities behind them, solid-wall insulation may be necessary which can decrease internal space and can be expensive to install. Energy-efficient windows are another important factor in insulation. Simply assuring that windows (and doors) are well sealed greatly reduces energy loss in a home. Double or Triple glazed windows are the typical method to insulating windows, trapping gas or creating a vacuum between two or three panes of glass allowing heat to be trapped inside or out. Low-emissivity or Low-E glass is another option for window insulation. It is a coating on windowpanes of a thin, transparent layer of metal oxide and works by reflecting heat back to its source, keeping the interior warm during the winter and cool during the summer. Simply hanging heavy-backed curtains in front of windows may also help their insulation. "Superwindows," mentioned in Natural Capitalism: Creating the Next Industrial Revolution, became available in the 1980s and use a combination of many available technologies, including two to three transparent low-e coatings, multiple panes of glass, and a heavy gas filling. Although more expensive, they are said to be able to insulate four and a half times better than a typical double-glazed windows.

Equipping roofs with highly reflective material (such as aluminium) increases a roof's albedo and will help reduce the amount of heat it absorbs, hence, the amount of energy needed to cool the building it is on. Green roofs or "living roofs" are a popular choice for thermally insulating a building. They are also popular for their ability to catch storm-water runoff and, when in the broader picture of a community, reduce the heat island effect thereby reducing energy costs of the entire area. It is arguable that they are able to replace the physical "footprint" that the building creates, helping reduce the adverse environmental impacts of the building's presence.

Energy efficiency and water conservation are also major considerations in sustainable housing. If using appliances, computers, HVAC systems, electronics, or lighting the sustainable-minded often look for an Energy Star label, which is government-backed and holds stricter regulations in energy and water efficiency than is required by law. Ideally, a sustainable shelter should be able to completely run the appliances it uses using renewable energy and should strive to have a neutral impact on the Earth's water sources

Greywater, including water from washing machines, sinks, showers and baths may be reused in landscape irrigation and toilets as a method of water conservation. Likewise, rainwater harvesting from storm-water runoff is also a sustainable method to conserve water usage in a sustainable shelter. Sustainable Urban Drainage Systems replicate the natural systems that clean water in wildlife and implement them in a city's drainage system so as to minimise contaminated water and unnatural rates of runoff into the environment.

Power

When needed, sustainable living requires the use of sustainable energy. This involves the use of power in such a way that fulfills the requirements of the present without compromising the requirements of the future. Or, in short, using power sources and in such a way that can be sustained infinitely. This means the energy source must be renewable, and must not harm the environment or the people working under it. The most commonly used renewable sources of energy are: biomass, water, geothermal, wind, and solar.

As mentioned under Shelter, some sustainable households may choose to produce their own renewable energy, while others may choose to purchase it through the grid from a power company that harnesses sustainable sources (also mentioned previously are the methods of metring the production and consumption of electricity in a household). Purchasing sustainable energy, however, may simply not be possible in some locations due to its limited availability. 6 out of the 50 states in the US do not offer green energy, for example.

For those that do, its consumers typically buy a fixed amount or a percentage of their monthly consumption from a company of their choice and the bought green energy is fed into the entire national grid. Technically, in this case, the green energy is not being fed directly to the household that buys it. In this case, it is possible that the actual amount of green electricity that the buying household receives is a small fraction of their total incoming electricity. This may or may not depend on the amount being purchased. The purpose of buying green electricity is to support their utility's effort in producing sustainable energy. Producing sustainable energy on an individual household or community basis is much more flexible, but can still be limited in the richness of the sources that the location may afford (some locations may not be rich in renewable energy sources while others may have an abundance of it).

When generating renewable energy and feeding it back into the grid (in participating countries such as the US and Germany), producing households are typically paid at least the full standard electricity rate by their utility and are also given separate renewable energy credits that they can then sell to their utility, additionally (utilities are interested in buying these renewable energy credits because it allows them to claim that they produce renewable energy). In some special cases, producing households may be paid up to four times the standard electricity rate, but this is not common.

Solar power harnesses the energy of the sun to make electricity. Two typical methods for converting solar energy into electricity are photo-voltaic cells that are organised into panels and concentrated solar power, which uses mirrors to concentrate sunlight to either heat a fluid that runs an electrical generator via a steam turbine or heat engine, or to simply cast onto photo-voltaic cells. The energy created by photo-voltaic cells is a direct current and has to be converted to alternating current before it can be used in a household. At this point, users can choose to either store this direct current in batteries for later use, or use an AC/DC inverter for immediate use. To get the best out of a solar panel, the angle of incidence of the sun should be between 20-50 degrees. Solor power via photo-voltaic cells are usually the most expensive method to harnessing renewable energy, but is falling in price as technology advances and public interest increases. It has the advantages of being portable, easy to use on an individual basis, readily available for government grants and incentives, and being flexible in regards to location (though it is most efficient when used in hot, arid areas since they tend to be the most sunny).

For those that are lucky, affordable rental schemes may be found. Concentrated solar power plants are typically used on more of a community scale rather than an individual household scale, because of the amount of energy they are able to harness but can be done on an individual scale with a parabolic reflector.

Solar thermal energy is harnessed by collecting direct heat from the sun. One of the most common ways that this method is used by households is through solar water heating. In a broad perspective, these systems involve well insulated tanks for storage and collectors, are either passive or active systems (active systems have pumps that continuously circulate water through the collectors and storage tank) and, in active systems, involve either directly heating the water that will be used or heating a non-freezing heat-transfer fluid that then heats the water that will be used. Passive systems are cheaper than active systems since they do not require a pumping system (instead, they take advantage of the natural movement of hot water rising above cold water to cycle the water being used through the collector and storage tank).

Other methods of harnessing solar power are solar space heating (for heating internal building spaces), solar drying (for drying wood chips, fruits, grains, etc.), solar cookers, solar distillers, and other passive solar technologies (simply, harnessing sunlight without any mechanical means).

Wind power is harnessed through turbines, set on tall towers (typically 20' or 6m with 10' or 3m diametre blades for an individual household's needs) that power a generator that creates electricity. They typically require an average of wind speed of 9 mi/hr (14 km/hr) to be worth their investment (as prescribed by the US Department of Energy), and are capable of paying for themselves within their lifetimes. Wind turbines in urban areas usually need to be mounted at least 30' (10m) in the air in order to receive enough wind and to be void of nearby obstructions (such as neighbouring buildings). Mounting a wind turbine may also require permission from authorities.

Wind turbines have been criticized for the noise they produce, their appearance, and the argument that they can affect the migratory patterns of birds (their blades obstruct passage in the sky).

Wind turbines are much more feasible for those living in rural areas and are one of the most cost-effective forms of renewable energy per kilowatt, approaching the cost of fossil fuels, and have quick paybacks.

For those that have a body of water flowing at an adequate speed (or falling from an adequate height) on their property, hydroelectricity may be an option. On a large scale, hydroelectricity, in the form of dams, has adverse environmental and social impacts. When on a small scale, however, in the form of single turbines, hydroelectricity is very sustainable. Single water turbines or even a group of single turbines are not environmentally or socially disruptive.

On an individual household basis, single turbines are the probably the only economically feasible route (but can have high paybacks and is one of the most efficient methods of renewable energy production). It is more common for an eco-village to use this method rather than a singular household.

Geothermal energy production involves harnessing the hot water or steam below the earth's surface, in reservoirs, to produce energy. Because the hot water or steam that is used is reinjected back into the reservoir, this source is considered sustainable.

However, those that plan on getting their electricity from this source should be aware that there is controversy over the lifespan of each geothermal reservoir as some believe that their lifespans are naturally limited (they cool down over time, making geothermal energy production there eventually impossible).

This method is often large scale as the system required to harness geothermal energy can be complex and requires deep drilling equipment. There do exist small individual scale geothermal operations, however, which harness reservoirs very close to the Earth's surface, avoiding the need for extensive drilling and sometimes even taking advantage of lakes or ponds where there is already a depression. In this case, the heat is captured and sent to a geothermal heat pump system located inside the shelter or facility that needs it (oftentimes, this heat is used directly to warm a greenhouse during the colder months).

Although geothermal energy is available everywhere on Earth, practicality and cost-effectiveness varies, directly related to the depth required to reach reservoirs. Places such as the Philippines, Hawaii, Alaska, Iceland, California, and Nevada have geothermal reservoirs closer to the Earth's surface, making its production cost-effective.

Biomass power is created when any biological matter is burned as fuel. As with the case of using green materials in a household, it is best to use as much locally available material as possible so as to reduce the carbon footprint created by transportation. Although burning biomass for fuel releases carbon dioxide, sulfur compounds, and nitrogen compounds into the atmosphere, a major concern in a sustainable lifestyle, the amount that is released is sustainable (it will not contribute to a rise in carbon dioxide levels in the atmosphere).

This is because the biological matter that is being burned releases the same amount of carbon dioxide that it consumed during its lifetime. However, burning biodiesel and bioethanol when created from virgin material, is increasingly controversial and may or may not be considered sustainable because it inadvertently increases global poverty, the clearing of more land for new agriculture fields (the source of the biofuel is also the same source of food), and may use unsustainable growing methods (such as the use of environmentally harmful pesticides and fertilizers).

List of organic matter than can be burned for fuel:

- Bagasse
- Biogas
- Manure
- Stover
- Straw
- Used vegetable oil
- Wood.

Digestion of organic material to produce methane is becoming an increasingly popular method of biomass energy production. Materials such as waste sludge can be digested to release methane gas that can then be burnt to produce electricity. Methane gas is also a natural by-product of landfills, full of decomposing waste, and can be harnessed here to produce electricity as well.

The advantage in burning methane gas is that is prevents the methane from being released into the atmosphere, exacerbating the greenhouse effect. Although this method of biomass energy production is typically large scale (done in landfills), it can be done on a smaller individual or community scale as well.

Food

Environmental Impacts of Industrial Agriculture

Industrial agricultural production is highly resource and energy intensive. Industrial agriculture systems typically require heavy irrigation, extensive pesticide and fertilizer application, intensive tillage, concentrated monoculture production, and other continual inputs. As a result of these industrial farming conditions, today's mounting environmental stresses are further exacerbated. These stresses include: declining water tables, chemical leaching, chemical runoff, soil erosion, land degradation, loss in biodiversity, and other ecological concerns.

Conventional Food Distribution and Long Distance Transport

Conventional food distribution and long distance transport is additionally resource and energy exhaustive. Substantial climate-disrupting carbon emissions, boosted by the transport of food over long distances, are of growing concern as the world faces such global crisis as natural resource depletion, peak oil and climate change. "The average American meal currently costs about 1500 miles, and takes about 10 calories of oil and other fossil fuels to produce a single calorie of food."

Local and Seasonal Foods

A more sustainable means of acquiring food is to purchase locally and seasonally. Buying food from local farmers reduces carbon offsets, caused by long-distance food transport, and stimulates the local economy. Local, small-scale farming operations also typically utilise more sustainable methods of agriculture than conventional industrial farming systems such as decreased tillage, nutrient cycling, fostered biodiversity and reduced chemical pesticide and fertilizer applications. Adapting a more regional, seasonally-based diet is more sustainable as it entails purchasing less energy and resource demanding produce that naturally grow within a local area and require no long-distance transport.

These vegetables and fruits are also grown and harvested within their suitable growing season. Thus, seasonal food farming does not require energy intensive greenhouse production, extensive irrigation, plastic packaging and long-distance transport from importing non-regional foods, and other environmental stressors. Local, seasonal produce is typically fresher, unprocessed and argued to be more nutritious. Local produce also contains less to no chemical residues from applications required for long-distance shipping and handling. Farmers' markets, public events where local small-scale farmers gather and sell their produce, are a good source for obtaining local food and knowledge about local farming productions. As well as promoting localization of food, farmers markets are a central gathering place for community interaction. Another way to become involved in regional food distribution is by joining a local community-supported agriculture (CSA). A CSA consists of a community of growers and consumers who pledge to support a farming operation while equally sharing the risks and benefits of food production. CSA's usually involve a system of weekly pick-ups of locally farmed vegetables and fruits, sometimes including dairy products, meat and special food items such as baked goods. Considering the previously noted rising environmental crisis, the United States and much of the world is facing immense vulnerability to famine. Local food production ensures food security if potential transportation disruptions and climatic, economical, and sociopolitical disasters were to occur.

Reducing Meat Consumption

Industrial meat production also involves high environmental costs such as land degradation, soil erosion and depletion of natural resources, especially pertaining to water and food. Buying and consuming organically-raised, free range or grass fed meat is another alternative towards more sustainable meat consumption.

Organic Farming

Purchasing and supporting organic products is another fundamental contribution to sustainable living. Organic farming is a rapidly emerging trend in the food industry and in the web of sustainability. According to the USDA National Organic Standards Board (NOSB), organic agriculture is defined as "an ecological production management system that promotes and enhances biodiversity, biological cycles, and soil biological activity. It is based on minimal use of off-farm inputs and on management practices that restore, maintain, or enhance ecological harmony. The primary goal of organic agriculture is to optimize the health and productivity of interdependent communities of soil life, plants, animals and people." Upon sustaining these goals, organic agriculture uses techniques such as crop rotation, permaculture, compost, green manure and biological pest control. In addition, organic farming prohibits or strictly limits the use of manufactured fertilizers and pesticides, plant growth regulators such as hormones, livestock antibiotics, food additives and genetically modified organisms. Organically farmed products include vegetables, fruit, grains, herbs, meat, dairy, eggs, fibres, and flowers.

Urban Gardening

In addition to local, small-scale farms, there has been a recent emergence in urban agriculture expanding from community gardens to private home gardens. With this trend, both farmers and ordinary people are becoming involved in food production. A network of urban farming systems helps to further ensure regional food security and encourages self-sufficiency and cooperative interdependence within communities. With every bite of food raised from urban gardens, negative environmental impacts are reduced in numerous ways. For instance, vegetables and fruits raised within small-scale gardens and farms are not grown with tremendous applications of nitrogen fertilizer required for industrial agricultural operations.

The nitrogen fertilizers cause toxic chemical leaching and runoff that enters our water tables. Nitrogen fertilizer also produces nitrous oxide, a more damaging greenhouse gas than carbon dioxide. Local, community-grown food also requires no imported, long-distance transport which further depletes our fossil fuel reserves. In developing more efficiency per land acre, urban gardens can be started in a wide variety of areas: in vacant lots, public parks, private yards, church and school yards, on roof tops (roof-top gardens), and many other places. Communities can work together in changing zoning limitations in order for public and private gardens to be permissible.

Aesthetically pleasing edible landscaping plants can also be incorporated into city landscaping such as blueberry bushes, grapevines trained on an arbor, pecan trees, etc. With as small a scale as home or community farming, sustainable and organic farming methods can easily be utilised. Such sustainable, organic farming techniques include: composting, biological pest control, crop rotation, mulching, drip irrigation, nutrient cycling and permaculture.

Food Preservation and Storage

Preserving and storing foods reduces reliance on long-distance transported food and the market industry. Home-grown foods can be preserved and stored outside of their growing season and continually consumed throughout the year, enhancing self-sufficiency and independence from the supermarket. Food can be preserved and saved by dehydration, freezing, vacuum packing, canning, bottling, pickling and jellying.

Transportation

With rising peak oil concerns, climate warming exacerbated by carbon emissions and high energy prices, the conventional automobile industry is becoming less and less feasible to the conversation of sustainability. Revisions of urban transport systems that foster mobility, low-cost transportation and healthier urban environments are needed. Such urban transport systems should consist of a combination of rail transport, bus transport, bicycle pathways and pedestrian walkways. Public transport systems such as underground rail systems and bus transit systems shift huge numbers of people away from reliance on car mobilisation and dramatically reduce the rate of carbon emissions caused by automobile transport. Carpooling is another alternative for reducing oil consumption and carbon emissions by transit.

In comparison with automobiles, bicycles are a paradigm of energy efficient personal transportation. Bicycles increase mobility while alleviating congestion, lowering air and noise pollution, and increasing physical exercise. Most importantly, they do not emit climate-disturbing carbon dioxide. Bike-sharing programs are beginning to boom throughout the world and are modelled in leading cities such as Paris, Amsterdam and London. Bike-sharing programs offer kiosks and docking stations that supply hundreds to thousands of bikes for rental throughout a city through small deposits or affordable memberships.

A recent boom has occurred in electric bikes especially in China and other Asian countries. Electric bikes are similar to plug-in hybrid vehicles in that they are battery powered and can be plugged into the provincial electric grid for recharging as needed. In contrast to plug-in hybrid cars, electric bikes do not directly use any fossil fuels. Adequate sustainable urban transportation is dependent upon proper city infrastructure and planning that incorporates efficient public transit along with bicycle and pedestrian-friendly pathways.

Water

A major factor of sustainable living involves that which no human can live without, water. Unsustainable water usage has far reaching implications for humankind. Currently, humans use one-fourth of the earth's total water in natural circulation, and over half the accessible runoff. Additionally, population growth and water demand is ever increasing. Thus, it is necessary to use available water more efficiently. In sustainable living, one can use water more sustainably through a series of simple, everyday measures. These measures involve considering indoor home appliance efficiency, outdoor water use, and daily water use awareness.

Indoor Home Appliances

Housing and commercial buildings account for 12 percent of America's freshwater withdrawals. A typical American single family home uses about 70 US gallons (260 L) per person per day indoors. This usage can be reduced by simple alterations in behaviour and upgrades to appliance quality.

Toilets

Toilets account for almost 30% of residential indoor water use in the United States. One flush of a standard US toilet requires more water than most individuals, and many families, in the world use for all their needs in an entire day. A home's toilet water sustainability can be improved in one of two ways: improving the current toilet or installing a more efficient toilet. To improve the current toilet, one possible method is to put weighted plastic bottles in the toilet tank. Also, there are inexpensive tank banks or float booster available for purchase. A tank bank is a plastic bag to be filled with water and hung in the toilet tank. A float booster attaches underneath the float ball of pre-1986 three and a half gallon capacity toilets. It allows these toilets to operate at the same valve and float setting but significantly reduces their water level, saving between one and one and a third gallons of water per flush. A major waste of water in existing toilets is leaks. A slow toilet leak is undetectable to the eye, but can waste hundreds of gallons each month. One way to check this is to put food dye in the tank, and to see if the water in the toilet bowl turns the same colour. In the event of a leaky flapper, one can replace it with an adjustable toilet flapper, which allows self adjustment of the amount of water per flush.

If installing a new toilet there are a number of options to obtain the most water efficient model. A low flush toilet uses one to two gallons per flush. Traditionally, toilets use three to five gallons per flush. If an eighteen litre per flush toilet is removed and a six litre per flush toilet is put in its place, 70% of the water flushed will be saved while the overall indoor water usage by will be reduced by 30%. It is possible to have a toilet that uses no water. A composting toilet treats human waste through composting and dehydration, producing a valuable soil additive. These toilets feature a two-compartment bowl to separate urine from feces. The urine can be collected or sold as fertilizer. The feces can be dried and bagged or composted. These toilets cost scarcely more than regularly installed toilets and do not require a sewer hookup. In addition to providing valuable fertilizer, these toilets are highly sustainable because they save sewage collection and treatment, as well as lessen agricultural costs and improve topsoil.

Additionally, one can reduce toilet water sustainability by limiting total toilet flushing. For instance, instead of flushing small wastes, such as tissues, one can dispose of these items using alternate measures.

Showers

On average, showers are 18% of indoor water use, at 6–8 US gallons (23–30 L) per minute traditionally in America. A simple method to reduce this usage is to switch to low-flow, high-performance showerheads. These showerheads use only 1.0-1.5 gpm or less. An alternative to replacing the showerhead is to install a converter. This device arrests a running shower upon reaching the desired temperature. Solar water heaters can be used to obtain optimal water temperature, and are more sustainable because they reduce dependence on fossil fuels. To lessen excess water usage, water pipes can be insulated with pre-slit foam pipe insulation. This insulation decreases hot water generation time. A simple, straightforward method to conserve water when showering is to take shorter showers. One method to accomplish this is to turn off the water when it is not necessary (such as while lathering) and resuming the shower when water is necessary.

Dishwasher/Sinks

On average, sinks are 15% of indoor water use. There are, however, easy methods to rectify excessive water loss. Available for purchase is a screw-on aerator. This device works by combining water with air thus generating a frothy substance that has more moisture and reduces water usage by half. Additionally, there is a flip-valve available that allows flow to be turned off and back on at the previously reached temperature. Finally, a laminar flow device creates a 1.5-2.4 gpm stream of water that reduces water usage by half, but can be turned to normal water level when optimal. In addition to buying the above devices, one can live more sustainably by checking sinks for leaks, and fixing these links if they exist. According to the EPA, "A small drip from a worn faucet washer can waste 20 gallons of water per day, while larger leaks can waste hundreds of gallons": When using a sink, being more aware of water usage is a very simple way to use water more sustainably. For instance, when washing dishes by hand, it is not necessary to leave the water running for rinsing. It is more sustainable to rinse dishes simultaneously. On average, dishwashing consumes 1% of indoor water use. When using a dishwasher, water can be conserved by only running the machine when it is completely full. Additionally, it can be set to Lowflow setting, in order to use less water per wash cycle. The enzymatic detergents available clean dishes more efficiently and more successfully with a smaller amount of water at a lower temperature.

Washing Machines

On average, 23% of indoor water use is due to clothes washing. In contrast to other machines, American washing machines have changed little to become more sustainable. A typical washing machine has a vertical-axis design, in which clothes are agitated in a tubful of water. Horizontal-axis machines, in contrast, put less water into the bottom of the rub and rotate clothes through it. These machines are more efficient in terms of soap usage and clothing stability.

Outdoor Water Usage

There are a number of ways one can incorporate a personal yard, roof, and garden in more sustainable living. While conserving water is a major element of sustainability, so is sequestering water.

Conserving Water

In planning a yard and garden space, it is most sustainable to consider the plants, soil, and available water. Drought resistant shrubs, plants, and grasses require a smaller amount of water in comparison to more traditional species. Additionally, native plants (as opposed to herbaceous perennials) will use a smaller supply of water and have a heightened resistance to plant diseases of the area. Xeriscape, a system that accounts for endemic features such as slope, soil type, and native plant range, can reduce landscape water use by 50 – 70%, while providing habitat space for wildlife. By planting slopes one can reduce runoff. Grouping plants by watering needs further reduces water waste.

After planting, placing a circumference of mulch surrounding plants functions to lessen evaporation. To do this, firmly press two to four inches of organic matter along the plant's dripline. This prevents water runoff. When watering, consider the range of sprinklers; watering paved areas is unnecessary. Additionally, to conserve the maximum amount of water, watering should be carried out during early mornings on non-windy days in order to reduce water loss to evaporation. Drip-irrigation systems and soaker hoses are a more sustainable alternative to the traditional sprinkler system. Drip-irrigation systems employ small gaps at standard distances in a hose, leading to the slow trickle of water droplets which percolate the soil over a protracted period. These systems use 30 – 50% less water than conventional methods. Soaker hoses help to reduce water use by up to 90%. They connect to a garden hose and lay along the row of plants under a layer of mulch. A layer of organic material added to the soil helps to increase its absorption and water retention; previously planted areas can be covered with compost.

In caring for a lawn, there are a number of measures that can increase the sustainability of lawn maintenance techniques. A primary aspect of lawn care is watering. In order to conserve water, it is important to only water when necessary, and to deep soak when watering. Additionally, a lawn may be left to go dormant, renewing after a dry spell to its original vitality.

Sequestering Water

A common method of water sequestrations is rainwater harvesting, which incorporates the collection and storage of rain. Primarily, the rain is obtained from a roof, and stored on the ground in catchment tanks. Water sequestration varies based on extent, cost, and complexity. A simple method involves a single barrel at the bottom of a downspout, while a more complex method involves multiple tanks. It is highly sustainable to use stored water in place of purified water for activities such as irrigation and flushing toilets. Additionally, using stored rainwater reduces the amount of runoff pollution, picked up from roofs and pavements that would normally enter streams through storm drains. The following equation can be used to estimate annual water supply:

Collection area (square feet) x Rainfall (inch/year) / 12 (inch/foot) = Cubic Feet of Water/Year.

Cubic Feet/Year x 7.43 (Gallons/Cubic Foot) = Gallons/year.

Note, however, this calculation does not account for losses such as evaporation or leakage.

Greywater systems function in sequestering used indoor water, such as laundry, bath and sink water, and filtering it for reuse. Greywater can be reused in irrigation and toilet flushing. There are two types of greywater systems: gravity fed manual systems and package systems. The manual systems do not require electricity but may require a larger yard space. The package systems require electricity but are self-contained and can be installed indoors.

Waste

As populations and resource demands climb, waste production contributes to emissions of carbon dioxide, leaching of hazardous materials into the soil and waterways, and methane emissions. In America alone, over the course of a decade, 500 trillion pounds of American resources will have been transformed into nonproductive wastes and gases. Thus, a crucial component of sustainable living is being waste conscious. One can do this by reducing waste, reusing commodities, and recycling.

There are a number of ways to reduce waste in sustainable living. One method is reducing paper waste, such as by taking action to cancel junk mail and move paper transactions to an online document. Another method to reduce waste is to buy in bulk, which reduces packaging materials. Preventing food waste is an alternative to organic waste compiling to create costly methane emissions. Food waste can be reintegrated into the environment through composting. Composting can be carried out at home or locally, with community composting. An additional example of how to reduce waste is being cognizant of not buying materials with limited use in excess, such as paint. Reduction aides in reducing the toxicity of waste if non-hazardous or less hazardous items are selected.

By reusing materials, one lives sustainably by not contributing to the addition of waste to landfills. Reuse saves natural resources by decreasing the necessity of raw material extraction. Recycling, a process that breaks down used items into raw materials in order to make new materials, is a particularly useful means of contributing to the renewal of goods. Recycling incorporates three primary processes; collection and processing, manufacturing, and purchasing recycled products. An offshoot of recycling, upcycling, strives to convert a material into something of similar or greater value in its second life. By integrating measures of reusing, reducing, and recycling one can effectively reduce production of waste and use materials in a sustainable manner.

E-Cycling

The term e-cycling is rather new, appearing on the USA EPA website which refers to donations, reuse, shredding and general collection of used electronics. Generically, the term refers to the process of collecting, brokering, disassembling, repairing or recycling the components or metals contained in used or discarded electronic equipment, otherwise known as electronic waste (e-waste). "E-cyclable" items include, but are not limited to: televisions, computers, microwave ovens, vacuum cleaners, telephones and cellular phones, stereos, and VCRs and DVDs. Investment in e-cycling facilities has been increasing recently due to technology's rapid rate of obsolescence, concern over improper methods, and opportunities for manufacturers to influence the secondary market (used and reused products). The controversy around methods stems from a lack of agreement over preferred outcomes. World markets with lower disposable incomes, for example, consider 75% repair and reuse to be valuable enough to justify 25% disposal. Regulated recyclers prefer 0% disposal, even if it means dramatically lower rates of reuse. Debate and certification standards may be leading to better definitions, though civil law contracts governing the expected process are still vital to any contracted process as poorly defined as "e-cycling".

Pros of e-cycling

Some people believe that any net disposal of e-waste following repair or metals recovery is unethical or illegal if it occurs in developing countries. Other people believe that the net environmental cost must include the mining, refining and extraction pollution costs of new product manufactured to replace secondary products which are destroyed in wealthy nations which cannot economically repair older product. As an example, groundwater has become so polluted in areas surrounding China's landfills that water must be shipped in from 18 miles away. However, mining of new metals has even broader impacts on groundwater.

Either e-cycling process, domestic processing or overseas repair, helps the environment by avoiding pollution and being a sustainable alternative to disposing of e-waste in landfills. Either domestic metals processing or overseas manual repair and e-cycling retrieved valuable raw materials from e-waste. Supporters of one form of "required e-cycling" legislation argue that e-cycling saves taxpayers money, as the financial responsibility would be shifted from the taxpayer to the manufacturers. Advocates of more simple legislation (such as landfill bans) argue that involving manufacturers does not reduce the cost to consumers, as reuse value is lost, and the resulting costs are passed on to consumers in new products (particularly affecting markets which cannot even afford those new products. It is theorized that manufacturers who take part in e-cycling will be motivated to use fewer materials in the production process, create longer lasting products, and implement safer, more efficient recycling systems.. This theory is sharply disputed and has never been demonstrated.

Criticisms of e-cycling

The critics of e-cycling are just as vocal as its advocates. According to the Reason Foundation, e-cycling will only raise the product and waste management costs of e-waste for consumers and limit innovation on the part of high-tech companies. They also believe that e-cycling facilities could unintentionally cause great harm to the environment. Additionally, critics claim that e-waste doesn't occupy a significant portion of total waste. According to a European study, only 4% of waste is electronic. Another opposition to e-cycling is that many problems are posed in disassembly: the process is costly and dangerous because of the heavy metals of which the electronic products are composed, and as little as 1-5% of the original cost of materials can be retrieved. A final problem that people find is that identity fraud is all too common in regards to the disposal of electronic products.. As the programs are legislated, creating winners and losers among e-cyclers with different locations and processes, it may be difficult to distinguish between criticism of ecycling as a practice and criticism of the specific legislated means proposed to enhance it.

Where Does e-waste Really Go?

A hefty criticism often lobbed at reuse based recyclers is that people think that they are recycling their electronic waste, when in reality it is actually being exported to developing countries such as China, India, and Nigeria. It has been estimated that 90% of e-waste is not being recycled as promised. (an article with no source for the statistic). For instance, at free recycling drives, "recyclers" may not be staying true to their word but are selling e-waste overseas or to parts brokers. Studies indicate that 50-80% of the 300,000-400,000 tons of e-waste is being sent overseas, and that approximately 2 million tons per year go to U.S. landfills. Although not possible in all circumstances, the best way to e-cycle is to upcycle your e-waste.. On the other hand, the electronic products in question are generally manufactured, and repaired under warranty, in the same nations which anti-reuse recyclers depict as primitive. Reuse-based erecyclers believe that fair-trade incentives for export markets will lead to better results than domestic shredding. The debate between export-friendly e-cycling and increased regulation of that practice was described in

What's Happening Now: Policy Issues and Current Efforts

Currently, pieces of government legislation and a number of grassroots efforts have contributed to the growth of e-cycling processes which emphasize decreased exports over increased reuse rates. The

Electronic Waste Recycling Act was passed in California in 2003. It requires that consumers pay an extra fee for certain types of electronics, and the collected money is then redistributed to recycling companies that are qualified to properly recycle these products.

It is the only state that legislates against e-waste through this kind of consumer fee, the other states' efforts focus on producer responsibility laws or waste disposal bans. No study has shown that per capita recovery is greater in one type of legislated program (e.g. California) vs. ordinary waste disposal bans (e.g. Massachusetts), though recovery is greatly increased in states which use either method.

As of September, 2006, Dell developed the nation's first completely free recycling program, furthering the responsibilities that manufacturers are taking for e-cycling. Additional manufacturers and retailers such as Best Buy, Sony, and Samsung have also set up recycling programs. This program does not accept televisions, which are the most expensive used electronic item, and are unpopular in markets which must deal with televisions when the more valuable computers have been cherry picked.

Another step being taken is the recyclers' pledge of true stewardship, sponsored by the Computer TakeBack Campaign. It has been signed by numerous recyclers promising to recycle responsibly. Grassroots efforts have also played a big part in this issue, as they and other community organisations are being formed to help responsibly recycle e-waste. Other grassroots campaigns are Basel, the Computer TakeBack Campaign (co-coordinated by the Grassroots Recycling Network), and the Silicon Valley Toxics Coalition. No study has shown any difference in recycling methods under the Pledge, and no data is available to demonstrate difference in management between "Pledge" and non-Pledge companies, though it is assumed that the risk of making false claims will prevent Pledge companies from wrongly describing their processes.

Many people believe that the U.S. should be following the European Union model in regards to its management of e-waste. In this program, a directive forces manufacturers to take responsibility for e-cycling; it also demands manufacturers' mandatory take-back and places bans on exporting e-waste to developing countries. Another longer-term solution is for computers to be composed of less dangerous products. Many people disagree. No data has been provided to show that people who agree with the European model have based their agreement on measured outcomes or experience-based scientific method.

Computer Recycling

Computer recycling or electronic recycling is the recycling or reuse of computers or other electronics. It includes both finding another use for materials (such as donation to charity), and having systems dismantled in a manner that allows for the safe extraction of the constituent materials for reuse in other products.

Reasons for Recycling

Obsolete computers or other electronics are a valuable source for secondary raw materials, if treated properly; if not treated properly, they are a source of toxins and carcinogens. Rapid technology change, low initial cost, and even planned obsolescence have resulted in a fast-growing surplus of computer or other electronic components around the globe. Technical solutions are available, but in most cases a legal framework, a collection system, logistics, and other services need to be implemented before a technical solution can be applied. According to the U.S. Environmental Protection Agency, an estimated 30 to 40 million surplus PCs, which it classifies under the term "hazardous household waste", will be ready for end-of-life management in each of the next few years. The U.S. National Safety Council estimates that 75% of all personal computers ever sold are now surplus electronics.

In 2007, the United States Environmental Protection Agency (EPA) said that more than 63 million computers in the U.S. were traded in for replacements—or they simply were discarded. Today 15 percent of electronic devices and equipment are recycled in the United States. Most electronic waste is sent to landfills or becomes incinerated, having a negative impact on the environment by releasing materials such as lead, mercury, or cadmium into the soil, groundwater, and atmosphere.

Many materials used in the construction of computer hardware can be recovered in the recycling process for use in future production. Reuse of tin, silicon, iron, aluminium, and a variety of plastics — all present in bulk in computers or other electronics — can reduce the costs of constructing new systems. In addition, components frequently contain copper, gold, and other materials valuable enough to reclaim in their own right.

Computer components contain valuable elements and substances suitable for reclamation, including lead, copper, and gold. They also contain many toxic substances, such as dioxins, polychlorinated biphenyls (PCBs), cadmium, chromium, radioactive isotopes, and mercury. A typical computer monitor may contain more than 6% lead by weight, much of which is in the lead glass of the cathode ray tube (CRT). A typical 15-inch computer monitor may contain 1.5 pounds of lead, but other monitors have been estimated as having up to 8 pounds of lead. Circuit boards contain considerable quantities of lead-tin solders and are even more likely to leach into groundwater or to create air pollution via incineration. Additionally, the processing required to reclaim the precious substances (including incineration and acid treatments) may release, generate, and synthesize further toxic byproducts.

A major computer or electronic recycling concern is export of waste to countries with lower environmental standards. Companies may find it cost-effective in the short term to sell outdated computers to less developed countries with lax regulations. It is commonly believed that a majority of surplus laptops are routed to developing nations as "dumping grounds for e-waste". The high value of working and reusable laptops, computers, and components (e.g., RAM) can help pay the cost of transportation for a large number of worthless "commodities". Broken monitors, obsolete circuit boards, and short-circuited transistors are difficult to spot in a containerload of used electronics.

Regulations

Europe

In Switzerland, the first electronic waste recycling system was implemented in 1991, beginning with collection of old refrigerators; over the years, all other electric and electronic devices were gradually added to the system. The established producer responsibility organisation is SWICO, mainly handling information, communication, and organisation technology.

The European Union implemented a similar system in February 2003, under the Waste Electrical and Electronic Equipment Directive (WEEE Directive, 2002/96/EC).

United States

Federal

The United States Congress considers a number of electronic waste bills, including the National Computer Recycling Act introduced by Congressman Mike Thompson (D-CA). Meanwhile, the main federal law governing solid waste is the Resource Conservation and Recovery Act of 1976. It covers only CRTs, though state regulations may differ. There are also separate laws concerning battery disposal. On March

25, 2009, the House Science and Technology Committee approved funding for research on reducing electronic waste and mitigating environmental impact, regarded by sponsor Ralph Hall (R-TX) as the first federal bill to address electronic waste directly.

State

Many states have introduced legislation concerning recycling and reuse of computers or computer parts or other electronics. Most American computer recycling legislation addresses it from within the larger electronic waste issue.

In 2001, Arkansas enacted the Arkansas Computer and Electronic Solid Waste Management Act, which requires that state agencies manage and sell surplus computer equipment, establishes a computer and electronics recycling fund, and authorizes the Department of Environmental Quality to regulate and/or ban the disposal of computer and electronic equipment in Arkansas landfills. The recently passed Electronic Device Recycling Research and Development Act distributes grants to universities, government labs, and private industry for research in developing projects in line with e-waste recycling and refurbishment.

Asia

South Korea, Japan, and Taiwan require that sellers and manufacturers of electronics be responsible for recycling 75% of them.

Recycling Methods

Consumer Recycling

Consumer recycling options include sale, donating computers directly to organisations in need, sending devices directly back to their original manufacturers, or getting components to a convenient recycler or refurbisher.

Corporate Recycling

Businesses seeking a cost-effective way to recycle large amounts of computer equipment responsibly face a more complicated process. Businesses also have the options of sale or contacting the Original Equipment Manufacturers (OEMs) and arranging recycling options.

Some companies will pick up unwanted equipment from businesses, wipe the data clean from the systems, and provide an estimate of the product's remaining value. For unwanted items that still have value, these firms will buy the excess IT hardware and sell refurbished products to those seeking more affordable options than buying new.

Companies that specialize in data protection and green disposal processes dispose of both data and used equipment while at the same time employing strict procedures to help improve the environment. Professional IT Asset Disposition (ITAD) firms specialize in corporate computer disposal and recycling services in compliance with local laws and regulations and also offer secure data elimination services that comply with data erasure standards.

Corporations face risks both for incompletely destroyed data and for improperly disposed computers. In America companies are liable for compliance with regulations even if the recycling process is outsourced under the Resource Conservation and Recovery Act. Companies can mitigate these risks by requiring waivers of liability, audit trails, certificates of data destruction, signed confidentiality agreements, and random audits of information security. The National Association of Information Destruction is an international trade association for data destruction providers.

Sale

Online auctions are an alternative for consumers willing to resell for cash less fees, in a complicated, self-managed, competitive environment where paid listings might not sell. Online classified ads can be similarly risky due to forgery scams and uncertainty.

Donation

A number of organisations attempt to reuse computers. These organisations usually refurbish usable computers for sale at discounted prices to schools, the needy, other nonprofit organisations, or the general public.

In the United States, consumer recycling includes a variety of donation options, such as charitable nonprofit organisations (NPOs) (501(c)(3) organisations - for example, Free Geek) which may offer tax benefits in return. NPOs (such as Nonprofit Technology Resources or Camara) will often accept and refurbish still-usable computers in return for tax benefits. The Computer Takeback Campaign and the TechSoup Donate Hardware List are resources for locating such refurbishers. Donated systems can also be directed to developing nations. However, in cases where the computer equipment comes from a wide variety of manufacturers, it may be more efficient to hire a third-party contractor (such as) to handle the recycling arrangements.

In Canada, Industry Canada runs the Computers for Schools program in association with TelecomPioneers.

Takeback

When researching computer companies before a computer purchase, consumers can find out if they offer recycling services. Most major computer manufacturers offer some form of recycling. At the user's request they may mail in their old computers, or arrange for pickup from the manufacturer.

Hewlett-Packard also offers free recycling, but only one of its "national" recycling programs is available nationally, rather than in one or two specific states. Hewlett-Packard also offers to pick up any computer product of any brand for a fee, and to offer a coupon against the purchase of future computers or components; it was the largest computer recycler in America in 2003, and it has recycled over 750 million pounds of electronic waste globally since 1995. It encourages the shared approach of collection points for consumers and recyclers to meet.

Exchange

Manufacturers often offer a free replacement service when purchasing a new PC. Dell Computers and Apple Inc. will take back old products when one buys a new one. Both refurbish and resell their own computers with a one-year warranty.

Many companies purchase and recycle all brands of working and broken laptops and notebook computers, whether from individuals or corporations. Building a market for recycling of desktop computers has proven more difficult than exchange programs for laptops, smartphones, and other smaller electronics. A basic business model is to provide a seller an instant online quote based on laptop characteristics, then to send a shipping label and prepaid box to the seller, to erase, reformat, and process the laptop, and to pay rapidly by check. A majority of these companies are also generalised electronic waste recyclers as well; organisations that recycle computers exclusively include Cash For Laptops, a laptop refurbisher in Nevada that claims to be the first to buy laptops online, in 2001.

Scrapping/Recycling

For systems which are obsolete or no longer useful to its user, recycling is often the only choice available. This is usually done by breaking down the equipment into its component parts, such as plastics and metals. These parts can then be recycled through various methods depending on the material. Recyclers typically charge a fee, but in return many have a zero-landfill policy and the sorted or shredded pieces are melted down to recover their component materials for re-use.

Early Pioneering Efforts to e-waste

The first major publication to report the recycling of computers and electronic waste was published on the front page of the New York Times on April 14, 1993 by columnist Steve Lohr.

Data Security

Data security is an important part of computer recycling. Federal regulations mandate that there are no information security leaks in the lifecycle of secure data; this includes its destruction and recycling. There are a number of federal laws and regulations, including HIPAA, Sarbanes-Oxley, FACTA, GLB, which govern the data lifecycle and require that establishments with high and low-profile data keep their data secure. Recycling computers can be dangerous when handling sensitive data, specifically to businesses storing tax records or employee information. While most people will try to wipe their hard drives clean before disposing of their old computers, only 5 percent rely on an industry specialist or a third party to completely clean the system before it's disposed of according to an IBM survey. Industry standards recommend a 3X overwriting process for complete protection against retrieving confidential information. This means a hard drive must be wiped three times in order to ensure the data cannot be retrieved and possibly used by others.

Reasons to Destroy and Recycle Securely

There are ways to ensure that not only hardware is destroyed but also the private data on the hard drive. Having customer data stolen, lost, or misplaced contributes to the ever growing number of people who are affected by identity theft, which can cause corporations to lose more than just money. The image of a company that holds secure data, such as banks, law firms, pharmaceuticals, and credit corporations is also at risk. If a company's public image is hurt that could cause consumers to not use their services and could cost millions in business losses and positive public relation campaigns.

The cost of data breaches "varies widely ranging $90 to $50,000 (under HIPAA's new HITECH amendment, that came about through the American Recovery and Revitalization act of 2009) per customer record, depending on whether the breach is "low-profile" or "high-profile" and the company is in a non-regulated or highly regulated area, such as banking or medical institutions." There is also a major backlash from the consumer if there is a data breach in a company that is supposed to be trusted to protect their private information. If an organisation has any consumer info on file, they must by law (Red Flags Clarification act of 2010) have written information protection policies and procedures in place, that serve to combat, mitigate, and detect vulnerable areas that could result in identity theft. The United States Department of Defence has published a standard to which recyclers and individuals may meet in order to satisfy HIPAA requirements.

Secure Recycling

There are regulations that monitor the data security on end-of-life hardware. National Association for Information Destruction (NAID) "is the international trade association for companies providing information destruction services. Suppliers of products, equipment and services to destruction companies are also eligible for membership. NAID's mission is to promote the information destruction industry and the standards and ethics of its member companies." There are companies that follow the guidelines from NAID and also meet all Federal EPA and local DEP regulations.

The typical process for computer recycling aims to securely destroy hard drives while still recycling the byproduct. A typical process for effective computer recycling accomplishes the following:

1. Receive hardware for destruction in locked and securely transported vehicles
2. Shred hard drives
3. Separate all aluminium from the waste metals with an electromagnet
4. Collect and securely deliver the shredded remains to an aluminium recycling plant
5. Mold the remaining hard drive parts into aluminium ingots.

Topics

- Data erasure
- Data remanence
- Degaussing
- Digger gold
- Electronic waste
- Global digital divide

- Green computing
- Polychlorinated biphenyls
- Trashware
- Found art.

Regulation

- Basel Convention
- Electronic Waste Recycling Act
- Electronic Waste Recycling Fee
- Material safety data sheet
- National Security Agency
- Restriction of Hazardous Substances Directive (RoHS)
- China RoHS
- Waste Electrical and Electronic Equipment Directive (WEEE directive).

Organisations

- ADISA
- Basel Action Network
- Blancco
- Camara
- Computer technology for developing areas
- CBL Data Recovery
- eDay
- Free Geek
- International Network for Environmental Compliance and Enforcement
- IT Asset Disposal Ltd
- Metech Incorporated (Metech Recycling)
- Tabernus
- Silicon Valley Toxics Coalition
- Solving the E-waste Problem
- Sustainable Electronics Initiative (SEI).

Digger Gold

Digger gold is the common slang term for gold recovered from electronics components such as board fingers, CPUs, and connector pins. For the gold fingers on boards or circuits, often a stripping solution is used to remove the gold from the board material, nitric acid also works well in this regard as many gold components are soldered to boards with silver-based solders that are soluble in nitric acid (which gold is not). After dissolving all other metals in solution, the digger gold is recovered by dissolution of the gold in aqua regia and subsequent selective precipitation of the gold using copperas or another selective reducing agent such as hydrazine. Due to the cost required and the small amount recovered, digger gold is not necessarily cost effective.

Electronic Waste by Country

Electronic waste is becoming an increasing part of the waste stream and efforts are being made to recycle and reduce this waste.

Basel Convention

The Basel Convention on the Control of Transboundary Movements of Hazardous Wastes and Their Disposal, usually known simply as the Basel Convention, is an international treaty that was designed to reduce the movements of hazardous waste between nations, and specifically to prevent transfer of hazardous waste from developed to less developed countries. Of the 172 parties to the Convention, Afghanistan, Haiti, and the United States have signed the Convention but have not yet ratified it.

Government Regulation

The United Nations Conference on Trade and Development (UNCTAD) tends to support the repair and recycling trade. Mining to produce the same metals, to meet demand for finished products in the west, also occurs in the same countries, and UNCTAD has recommended that restrictions against recycling exports be balanced against the environmental costs of recovering those materials from mining. Hard rock mining produces 45% of all toxins produced by all industries in the United States.

Greenpeace contends that residue problems are so significant that the exports of all used electronics should be banned.

Asia

Many Asian countries have legislated, or will do so, for electronic waste recycling. South Korea, Japan and Taiwan ensure manufacturer responsibility by demanding that they recycle 75% of their annual production.

China

Chinese laws are primarily concerned with eliminating the import of e-waste. China has ratified the Basel Convention as well as the Basel Ban Amendment, officially banning the import of e-waste. In October 2008, The Chinese State Council also approved a "draft regulation on the management of electronic waste." This regulation is intended to promote the continued use of resources through recycling and to monitor the end-of-life treatment of electronics. Under the new regulations, recycling of electronics by the consumer is mandated. It also requires the recycling of unnecessary materials discarded in the manufacturing process.

Japan

Europe

Some European countries implemented laws prohibiting the disposal of electronic waste in landfills in the 1990s. "This created an e-waste processing industry in Europe."

In Switzerland, the first electronic waste recycling system was implemented in 1991, beginning with collection of old refrigerators. Over the years, all other electric and electronic devices were gradually been included in the system. Legislation followed in 1998, and since January 2005 it has been possible to return all electronic waste to the sales points and other collection points free of charge. There are two established producer responsibility organisations: SWICO, mainly handling information, communication, and organisation technology, and SENS, responsible for electrical appliances. The total amount of recycled electronic waste exceeds 10 kg per capita per year.

Additionally, the European Union has implemented several directives and regulations that place the responsibility for "recovery, reuse and recycling" on the manufacturer.

The Waste Electrical and Electronic Equipment Directive (WEEE Directive), as it is often referred to, has now been transposed in national laws in all member countries of the European Union. It was designed to make equipment manufacturers financially or physically responsible for their equipment at the end of its life, under a policy known as Extended producer responsibility (EPR). "Users of electrical and electronic equipment from private households should have the possibility of returning WEEE at least free of charge", and manufacturers must dispose of it in an environmentally friendly manner, by ecological disposal, reuse, or refurbishment. EPR is seen as a useful policy as it internalizes the end-of-life costs and provided a competitive incentive for companies to design equipment with fewer costs and liabilities when it reached its end of life. However, the application of the WEEE Directive has been criticized for implementing the EPR concept in a collective manner, and thereby losing the competitive incentive of individual manufacturers to be rewarded for their green design. Since August 13, 2005, electronics manufacturers have become financially responsible for compliance to the WEEE Directive.

Under the directive, each country recycles at least 4 kg of electronic waste per capita per year. Furthermore, the Directive should "decrease e-waste and e-waste exports.". In December 2008 a draft revision to the Directive proposed a market-based goal of 65%, which is 22 kg per capita in the case of the United Kingdom. A decision on the proposed revisions could result in a new WEEE Directive by 2012.

The Directive on the Restriction of the Use of Certain Hazardous Substances in Electrical and Electronic Equipment (2002/95/EC), commonly referred to as the Restriction of Hazardous Substances Directive (RoHS Directive), was also adopted in February 2003 by the European Union. The RoHS Directive took effect on July 1, 2006, and is required to be enforced and become law in each member state. This directive restricts the use of six hazardous materials in the manufacture of various types of electronic and electrical equipment.

The Battery Directive enacted in 2006 regulates the manufacture, disposal and trade of batteries in the European Union.

North America

Canada

In February 2004, a fee similar to the one in California was added to the cost of purchasing new televisions, computers, and computer components in Alberta, the first of its kind in Canada. Saskatchewan also implemented an electronics recycling fee in February 2007, followed by British Columbia in August 2007, Nova Scotia in February 2008, and Ontario in April 2009. In 2007, Manitoba issued the Proposed Electrical and Electronic Equipment Stewardship Regulation by which the sale of regulated products is forbidden unless covered by the stewardship program. "Products covered under this legislation include TVs, computers, laptops, and scanners." Recycling regulation passed in Ontario in October 2004, requires producers to "either develop product stewardship plans or comply with a product stewardship program for specific products."

Canadian Federal Legislation

The Export and Import of Hazardous Waste and Hazardous Recyclable Material Regulations (EIHWHRMR) operates with a few basic premises, one of which being that electronic waste is either "intact" or "not intact". The various annexes define hazardous waste in Canada, and also deem any waste that is "...considered or defined as hazardous under the legislation of the country receiving it and is prohibited by that country from being imported or conveyed in transit" to be covered under Canadian regulation and therefore subject to prior informed consent procedures.

The loophole in the regulations that allows tons of e-waste to be exported from Canada is the use of the definition of "intact" vs "functional". A non-functioning electronic device that is intact can be exported under the current legislation. What can't be exported is a non-functioning but no longer intact electronic device. The principal problem being, the non-functioning electronic device is at high risk of being disassembled in some far away e-waste dumping ground. The Canadian government's use of a unique interpretation of the Basel Convention obligations "intact" and "not intact" opens the door to uncontrolled e-waste exports as long as the device is intact.

Since Canada ratified the Basel Convention on August 28, 1992, and as of August 2011, Environment Canada's Enforcement Branch has initiated 176 investigations for violations under EIHWHRMR, some of which are still in progress. There have been 19 prosecutions undertaken for non-compliance with the provisions of the EIHWHRMR some of which are still before the courts.

Oceania

Australia

Electronic waste has been on the agenda of the Australian Federal Government since the mid 1990s. The Australian and New Zealand Environment and Conservation Council (now replaced by the Environment Protection and Heritage Council (EPHC)) was the first body to identify electrical and electronic waste as a concern. In 2002, the EPHC again declared that e-waste needed action. The Electrical Equipment Product Stewardship Sub-Group examined the issue and decided that computer and television waste were 'wastes of concern'. Since that time the television and computer industry has been working with the EPHC to identify a suitable way to manage end-of-life televisions and computers.

In November 2008 the EPHC committed to the development of a national solution to the issue of managing television and computer waste. This action culminated in the release of a package of documents designed to enable public consultation on the various options for managing end-of-life televisions and computers on 16 July 2009. The main document in the package is the Consultation Regulatory Impact Statement: Televisions and Computers. The paper canvasses various options for managing end-of-life units and analyses the costs and benefits of each. The Consultation Paper does not have a preferred option. The preferred option will be developed by government through the public consultation process prior to the next meeting of the EPHC on 5 November 2009 in Perth where State and Federal Minister will adopt a position.

A series of public meetings were held in Adelaide, Perth, Sydney and Melbourne to receive feedback to the government's proposals. The meetings occurred in late July and early August 2009.

Product Stewardship

Product Stewardship Australia (PSA)is a not-for-profit organisation established by the television industry in Australia to lead the way in developing recycling programs for e-waste in Australia, particularly televisions. PSA works closely with both State and Federal Governments along with other industry associations to advance product stewardship in Australia. PSA has contributed to the development of the Consultation Regulatory Impact Statement on Televisions and Computers.

Green Computing

Green computing or green IT, refers to environmentally sustainable computing or IT. In the article *Harnessing Green IT: Principles and Practices*, San Murugesan defines the field of green computing as "the study and practice of designing, manufacturing, using, and disposing of computers, servers, and associated subsystems—such as monitors, printers, storage devices, and networking and communications systems — efficiently and effectively with minimal or no impact on the environment." The goals of green computing are similar to green chemistry; reduce the use of hazardous materials, maximise energy efficiency during the product's lifetime, and promote the recyclability or biodegradability of defunct products and factory waste. Research continues into key areas such as making the use of computers as energy-efficient as possible, and designing algorithms and systems for efficiency-related computer technologies.

Origins

In 1992, the U.S. Environmental Protection Agency launched Energy Star, a voluntary labelling program that is designed to promote and recognise energy-efficiency in monitors, climate control equipment, and other technologies. This resulted in the widespread adoption of sleep mode among consumer electronics. The term "green computing" was probably coined shortly after the Energy Star program began; there are several USENET posts dating back to 1992 that use the term in this manner.

Concurrently, the Swedish organisation TCO Development launched the TCO Certification program to promote low magnetic and electrical emissions from CRT-based computer displays; this program was later expanded to include criteria on energy consumption, ergonomics, and the use of hazardous materials in construction.

Regulations and Industry Initiatives

The Organisation for Economic Co-operation and Development (OECD) has published a survey of over 90 government and industry initiatives on "Green ICTs", i.e. information and communication technologies, the environment and climate change. The report concludes that initiatives tend to concentrate on the greening ICTs themselves rather than on their actual implementation to tackle global warming and environmental degradation. In general, only 20% of initiatives have measurable targets, with government programs tending to include targets more frequently than business associations.

Government

Many governmental agencies have continued to implement standards and regulations that encourage green computing. The Energy Star program was revised in October 2006 to include stricter efficiency requirements for computer equipment, along with a tiered ranking system for approved products.

Some efforts place responsibility on the manufacturer to dispose of the equipment themselves after it is no longer needed; this is called the extended producer responsibility model. The European Union's directives 2002/95/EC (Restriction of Hazardous Substances Directive), on the reduction of hazardous substances, and 2002/96/EC (Waste Electrical and Electronic Equipment Directive) on waste electrical and electronic equipment required the substitution of heavy metals and flame retardants like Polybrominated biphenyl and Polybrominated diphenyl ethers in all electronic equipment put on the market starting on July 1, 2006. The directives placed responsibility on manufacturers for the gathering and recycling of old equipment.

There are currently 26 US states that have established state-wide recycling programs for obsolete computers and consumer electronics equipment. The statutes either impose an "advance recovery fee" for each unit sold at retail or require the manufacturers to reclaim the equipment at disposal.

In 2010, the American Recovery and Reinvestment Act (ARRA) was signed into legislation by President Obama. The bill allocated over $90 billion to be invested in green initiatives (renewable energy, smart grids, energy efficiency, etc.) In January 2010, the U.S. Energy Department granted $47 million of the ARRA money towards projects that aim to improve the energy efficiency of data centres. The projects will provide research on the following three areas: optimize data centre hardware and software, improve power supply chain, and data centre cooling technologies.

Industry

- Climate Savers Computing Initiative (CSCI) is an effort to reduce the electric power consumption of PCs in active and inactive states. The CSCI provides a catalog of green products from its member organisations, and information for reducing PC power consumption. It was started on 2007-06-12. The name stems from the World Wildlife Fund's Climate Savers program, which was launched in 1999. The WWF is also a member of the Computing Initiative.
- The Green Electronics Council offers the Electronic Product Environmental Assessment Tool (EPEAT) to assist in the purchase of "greener" computing systems. The Council evaluates computing equipment on 51 criteria - 23 required and 28 optional - that measure a product's efficiency and sustainability attributes. Products are rated Gold, Silver, or Bronze, depending on how many optional criteria they meet. On 2007-01-24, President George W. Bush issued Executive Order 13423, which requires all United States Federal agencies to use EPEAT when purchasing computer systems.
- The Green Grid is a global consortium dedicated to advancing energy efficiency in data centres and business computing ecosystems. It was founded in February 2007 by several key companies in the industry – AMD, APC, Dell, HP, IBM, Intel, Microsoft, Rackable Systems, SprayCool, Sun Microsystems

and VMware. The Green Grid has since grown to hundreds of members, including end-users and government organisations, all focused on improving data centre infrastructure efficiency (DCIE).

- The Green500 list rates supercomputers by energy efficiency (megaflops/watt, encouraging a focus on efficiency rather than absolute performance.
- Green Comm Challenge is an organisation that promotes the development of energy conservation technology and practices in the field of Information and Communications Technology (ICT).
- The Transaction Processing Performance Council(TPC) Energy specification augments the existing TPC benchmarks by allowing for optional publications of energy metrics alongside their performance results.
- The SPEC Power is the first industry standard benchmark that measures power consumption in relation to performance for server-class computers.

Approaches

In the article *Harnessing Green IT: Principles and Practices*, San Murugesan defines the field of green computing as "the study and practice of designing, manufacturing, using, and disposing of computers, servers, and associated subsystems — such as monitors, printers, storage devices, and networking and communications systems — efficiently and effectively with minimal or no impact on the environment." Murugesan lays out four paths along which he believes the environmental effects of computing should be addressed: Green use, green disposal, green design, and green manufacturing.

Green computing can also develop solutions that offer benefits by "aligning all IT processes and practices with the core principles of sustainability, which are to reduce, reuse, and recycle; and finding innovative ways to use IT in business processes to deliver sustainability benefits across the enterprise and beyond".

Modern IT systems rely upon a complicated mix of people, networks, and hardware; as such, a green computing initiative must cover all of these areas as well. A solution may also need to address end user satisfaction, management restructuring, regulatory compliance, and return on investment (ROI). There are also considerable fiscal motivations for companies to take control of their own power consumption; "of the power management tools available, one of the most powerful may still be simple, plain, common sense."

Product Longevity

Gartner maintains that the PC manufacturing process accounts for 70 % of the natural resources used in the life cycle of a PC. More recently, Fujitsu released a Life Cycle Assessment (LCA) of a desktop that show that manufacturing and end of life accounts for the majority of this laptop ecological footprint. Therefore, the biggest contribution to green computing usually is to prolong the equipment's lifetime. Another report from Gartner recommends to "Look for product longevity, including upgradability and modularity." For instance, manufacturing a new PC makes a far bigger ecological footprint than manufacturing a new RAM module to upgrade an existing one.

Software and Deployment Optimization

Algorithmic Efficiency

The efficiency of algorithms has an impact on the amount of computer resources required for any given computing function and there are many efficiency trade-offs in writing programs. While algorithmic efficiency does not have as much impact as other approaches, it is still an important consideration. A study by a physicist at Harvard, estimated that the average Google search released 7 grams of carbon dioxide (CO_2). However, Google disputes this figure, arguing instead that a typical search produces only 0.2 grams of CO_2. More recently, an independent study by GreenIT.fr demonstrate that Windows 7 + Office 2010 require 70 times more memory (RAM) than Windows 98 + Office 2000 to write exactly the same text or send exactly the same

Resource Allocation

Algorithms can also be used to route data to data centres where electricity is less expensive. Researchers from MIT, Carnegie Mellon University, and Akamai have tested an energy allocation algorithm that successfully routes traffic to the location with the cheapest energy costs. The researchers project up to a 40 percent savings on energy costs if their proposed algorithm were to be deployed. However, this approach does not actually reduce the amount of energy being used; it reduces only the cost to the company using it. Nonetheless, a similar strategy could be used to direct traffic to rely on energy that is produced in a more environmentally friendly or efficient way. A similar approach has also been used to cut energy usage by routing traffic away from data centres experiencing warm weather; this allows computers to be shut down to avoid using air conditioning.

Larger server centres are sometimes located where energy and land are inexpensive and readily available. Local availability of renewable energy, climate that allows outside air to be used for cooling, or locating them where the heat they produce may be used for other purposes could be factors in green siting decisions.

Virtualization

Computer virtualization refers to the abstraction of computer resources, such as the process of running two or more logical computer systems on one set of physical hardware. The concept originated with the IBM mainframe operating systems of the 1960s, but was commercialized for x86-compatible computers only in the 1990s. With virtualization, a system administrator could combine several physical systems into virtual machines on one single, powerful system, thereby unplugging the original hardware and reducing power and cooling consumption. Virtualization can assist in distributing work so that servers are either busy or put in a low-power sleep state. Several commercial companies and open-source projects now offer software packages to enable a transition to virtual computing. Intel Corporation and AMD have also built proprietary virtualization enhancements to the x86 instruction set into each of their CPU product lines, in order to facilitate virtualized computing.

Terminal Servers

Terminal servers have also been used in green computing. When using the system, users at a terminal connect to a central server; all of the actual computing is done on the server, but the end user experiences the operating system on the terminal. These can be combined with thin clients, which use up to 1/8 the amount of energy of a normal workstation, resulting in a decrease of energy costs and consumption. There has been an increase in using terminal services with thin clients to create virtual labs. Examples of terminal server software include Terminal Services for Windows and the Linux Terminal Server Project (LTSP) for the Linux operating system.

Power Management

The Advanced Configuration and Power Interface (ACPI), an open industry standard, allows an operating system to directly control the power-saving aspects of its underlying hardware. This allows a system to automatically turn off components such as monitors and hard drives after set periods of inactivity. In addition, a system may hibernate, where most components (including the CPU and the system RAM) are turned off. ACPI is a successor to an earlier Intel-Microsoft standard called Advanced Power Management, which allows a computer's BIOS to control power management functions.

Some programs allow the user to manually adjust the voltages supplied to the CPU, which reduces both the amount of heat produced and electricity consumed. This process is called undervolting. Some CPUs can automatically undervolt the processor, depending on the workload; this technology is called "SpeedStep" on Intel processors, "PowerNow!"/"Cool'n'Quiet" on AMD chips, LongHaul on VIA CPUs, and LongRun with Transmeta processors.

Data Centre Power

Data centres, which have been criticized for its extraordinary high energy demand, are a primary focus for proponents of green computing. The federal government has set a minimum 10% reduction target for data centre energy usage by 2011. With the aid of a self-styled ultraefficient evaporative cooling technology.

Operating System Support

The dominant desktop operating system, Microsoft Windows, has included limited PC power management features since Windows 95. These initially provided for stand-by (suspend-to-RAM) and a monitor low power state. Further iterations of Windows added hibernate (suspend-to-disk) and support for the ACPI standard. Windows 2000 was the first NT-based operating system to include power management. This required major changes to the underlying operating system architecture and a new hardware driver model. Windows 2000 also introduced Group Policy, a technology that allowed administrators to centrally configure most Windows features.

However, power management was not one of those features. This is probably because the power management settings design relied upon a connected set of per-user and per-machine binary registry values, effectively leaving it up to each user to configure their own power management settings.

This approach, which is not compatible with Windows Group Policy, was repeated in Windows XP. The reasons for this design decision by Microsoft are not known, and it has resulted in heavy criticism. Microsoft significantly improved this in Windows Vista by redesigning the power management system to allow basic configuration by Group Policy. The support offered is limited to a single per-computer policy. The most recent release, Windows 7 retains these limitations but does include refinements for more efficient user of operating system timers, processor power management, and display panel brightness. The most significant change in Windows 7 is in the user experience. The prominence of the default High Performance power plan has been reduced with the aim of encouraging users to save power.

There is a significant market in third-party PC power management software offering features beyond those present in the Windows operating system available. Most products offer Active Directory integration and per-user/per-machine settings with the more advanced offering multiple power plans, scheduled power plans, anti-insomnia features and enterprise power usage reporting. Notable vendors include 1E NightWatchman., Data Synergy PowerMAN (Software), Faronics Power Save and Verdiem SURVEYOR.

Power Supply

Desktop computer power supplies (PSUs) are in general 70–75% efficient, dissipating the remaining energy as heat. An industry initiative called 80 PLUS certifies PSUs that are at least 80% efficient; typically these models are drop-in replacements for older, less efficient PSUs of the same form factor. As of July 20, 2007, all new Energy Star 4.0-certified desktop PSUs must be at least 80% efficient.

Storage

Smaller form factor (e.g., 2.5 inch) hard disk drives often consume less power per gigabyte than physically larger drives. Unlike hard disk drives, solid-state drives store data in flash memory or DRAM. With no moving parts, power consumption may be reduced somewhat for low-capacity flash-based devices.

In a recent case study, Fusion-io, manufacturers of the world's fastest Solid State Storage devices, managed to reduce the carbon footprint and operating costs of MySpace data centres by 80% while increasing performance speeds beyond that which had been attainable via multiple hard disk drives in Raid 0. In response, MySpace was able to permanently retire several of their servers, including all their heavy-load servers, further reducing their carbon footprint.

As hard drive prices have fallen, storage farms have tended to increase in capacity to make more data available online. This includes archival and backup data that would formerly have been saved on tape or other offline storage. The increase in online storage has increased power consumption. Reducing the power consumed by large storage arrays, while still providing the benefits of online storage, is a subject of ongoing research.

Video Card

A fast GPU may be the largest power consumer in a computer. Energy-efficient display options include:

- No video card - use a shared terminal, shared thin client, or desktop sharing software if display required.
- Use motherboard video output - typically low 3D performance and low power.
- Select a GPU based on low idle power, average wattage, or performance per watt.

Display

CRT monitors typically use more power than LCD monitors. They also contain significant amounts of lead. LCD monitors typically use a cold-cathode fluorescent bulb to provide light for the display. Some newer displays use an array of light-emitting diodes (LEDs) in place of the fluorescent bulb, which reduces the amount of electricity used by the display. Fluorescent back-lights also contain mercury, whereas LED back-lights do not.

Materials Recycling

Recycling computing equipment can keep harmful materials such as lead, mercury, and hexavalent chromium out of landfills, and can also replace equipment that otherwise would need to be manufactured, saving further energy and emissions. Computer systems that have outlived their particular function can be re-purposed, or donated to various charities and non-profit organisations.

However, many charities have recently imposed minimum system requirements for donated equipment. Additionally, parts from outdated systems may be salvaged and recycled through certain retail outlets and municipal or private recycling centres. Computing supplies, such as printer cartridges, paper, and batteries may be recycled as well.

A drawback to many of these schemes is that computers gathered through recycling drives are often shipped to developing countries where environmental standards are less strict than in North America and Europe. The Silicon Valley Toxics Coalition estimates that 80%

of the post-consumer e-waste collected for recycling is shipped abroad to countries such as China and Pakistan.

In 2011, the collection rate of e-waste is still very low, even in the most ecology-responsible countries like France. In this country, e-waste collection is still at a 14% annual rate between electronic equipments sold and e-waste collected for 2006 to 2009.

The recycling of old computers raises an important privacy issue. The old storage devices still hold private information, such as emails, passwords, and credit card numbers, which can be recovered simply by someone's using software available freely on the Internet. Deletion of a file does not actually remove the file from the hard drive.

Before recycling a computer, users should remove the hard drive, or hard drives if there is more than one, and physically destroy it or store it somewhere safe. There are some authorized hardware recycling companies to whom the computer may be given for recycling, and they typically sign a non-disclosure agreement.

Telecommuting

Teleconferencing and telepresence technologies are often implemented in green computing initiatives. The advantages are many; increased worker satisfaction, reduction of greenhouse gas emissions related to travel, and increased profit margins as a result of lower overhead costs for office space, heat, lighting, etc. The savings are significant; the average annual energy consumption for U.S. office buildings is over 23 kilowatt hours per square foot, with heat, air conditioning and lighting accounting for 70% of all energy consumed. Other related initiatives, such as hotelling, reduce the square footage per employee as workers reserve space only when they need it. Many types of jobs, such as sales, consulting, and field service, integrate well with this technique.

Voice over IP (VoIP) reduces the telephony wiring infrastructure by sharing the existing Ethernet copper. VoIP and phone extension mobility also made hot desking more practical.

Education and Certification

Green Computing Degree Programs

Degree programs that provide training in a range of information technology concentrations along with sustainable strategies in an effort to educate students how to build and maintain systems while reducing its negative impact on the environment.

Green Computing Certifications

Some certifications demonstrate that an individual has specific green computing knowledge, including:

- Green Computing Initiative - CGCUS, CGCA, CGCP certifications
- CompTIA Strata Green IT is designed for IT managers to show that they have good knowledge of green IT practices and methods and why it is important to incorporate them into an organisation.
- Information Systems Examination Board (ISEB) Foundation Certificate in Green IT is appropriate for showing an overall understanding and awareness of green computing and where its implementation can be beneficial.
- Singapore Infocomm Technology Federation (SiTF) Singapore Green IT Professional is industry endorsed professional level certification which uses a Green IT framework (adopted by over 13 countries) and is inclusive of vendors which provides a in depth understanding of Green IT and the implementation of these practices.

Camara (Charity) (Ireland)

- *Challenging the Chip*, a book about labour rights and environmental justice in the global electronics industry
- Desktop virtualization
- Data migration
- Digger gold
- e-cycling
- eDay, an electronic waste collection day in New Zealand
- Electronic Waste Recycling Act
- Energy Efficient Ethernet
- Energy consumption of computers in the USA
- Interconnect bottleneck
- IT energy management
- Minimalism (computing)
- Optical communication
- Optical fibre cable

- Optical interconnect
- Parallel optical interface
- Plug computer
- Power factor
- Power usage effectiveness (PUE)
- Rebound effect (paradoxical negative effect)
- Restriction of Hazardous Substances Directive (RoHS)
- Standby power
- Sustainable Electronics Initiative (SEI)
- Time-sharing
- Thunderbolt (interface)
- Trashware
- Virtual application.

Chapter 5

Restriction of Hazardous Substances Directive

The Directive on the restriction of the use of certain hazardous substances in electrical and electronic equipment 2002/95/EC; commonly referred to as the Restriction of Hazardous Substances Directive or RoHS) was adopted in February 2003 by the European Union. The RoHS directive took effect on 1 July 2006, and is required to be enforced and become law in each member state. This directive restricts the use of six hazardous materials in the manufacture of various types of electronic and electrical equipment. It is closely linked with the Waste Electrical and Electronic Equipment Directive (WEEE) 2002/96/EC which sets collection, recycling and recovery targets for electrical goods and is part of a legislative initiative to solve the problem of huge amounts of toxic e-waste.

Details

Each European Union member state will adopt its own enforcement and implementation policies using the directive as a guide.

RoHS is often referred to as the lead-free directive, but it restricts the use of the following six substances:

1. Lead (Pb)
2. Mercury (Hg)
3. Cadmium (Cd)
4. Hexavalent chromium (Cr^{6+})
5. Polybrominated biphenyls (PBB)
6. Polybrominated diphenyl ether (PBDE).

PBB and PBDE are flame retardants used in several plastics.

The maximum permitted concentrations are 0.1% or 1000 ppm (except for cadmium, which is limited to 0.01% or 100 ppm) by weight of *homogeneous material*. This means that the limits do not apply to the weight of the finished product, or even to a component, but to any single substance that could (theoretically) be separated mechanically—for example, the sheath on a cable or the tinning on a component lead.

As an example, a radio is composed of a case, screws, washers, a circuit board, speakers, etc. The screws, washers, and case may each be made of homogenous materials, but the other components comprise multiple sub-components of many different types of material. For instance, a circuit board is composed of a bare PCB, ICs, resistors, capacitors, switches, etc. A switch is composed of a case, a lever, a spring, contacts, pins, etc., each of which may be made of different materials. A contact might be composed of a copper strip with a surface coating. A speaker is composed of a permanent magnet, copper wire, paper, etc.

Everything that can be identified as a homogeneous material must meet the limit. So if it turns out that the case was made of plastic with 2,300 ppm (0.23%) PBB used as a flame retardant, then the entire radio would fail the requirements of the directive.

In an effort to close RoHS loopholes, in May 2006 the European Commission was asked to review two currently excluded product categories (monitoring and control equipment, and medical devices) for future inclusion in the products that must fall into RoHS compliance. In addition the commission entertains requests for deadline extensions or for exclusions by substance categories, substance location or weight. New legislation was published in the official journal in July, 2011 which supersedes this exemption.

Note that batteries are not included within the scope of RoHS. However, in Europe, batteries are under the European Commission's 1991 Battery Directive (91/157/EEC), which was recently increased in scope and approved in the form of the new battery directive, version 2003/0282 COD, which will be official when submitted to and published in the EU's Official Journal.

While the first Battery Directive addressed possible trade barrier issues brought about by disparate European member states' implementation, the new directive more explicitly highlights improving and protecting the environment from the negative effects of the waste contained in batteries. It also contains a program for more ambitious recycling of industrial, automotive, and consumer batteries, gradually increasing the rate of manufacturer-provided collection sites to 45% by 2016. It also sets limits of 5 ppm mercury and 20 ppm cadmium to batteries except those used in medical, emergency, or portable power-tool devices. Though not setting quantitative limits on quantities of lead, lead-acid, nickel, and nickel-cadmium in batteries, it cites a need to restrict these substances and provide for recycling up to 75% of batteries with these substances. There are also provisions for marking the batteries with symbols in regard to metal content and recycling collection information.

The directive applies to equipment as defined by a section of the WEEE directive. The following numeric categories apply:

1. Large household appliances.
2. Small household appliances.
3. IT & Telecommunications equipment (although infrastructure equipment is exempt in some countries)
4. Consumer equipment.
5. Lighting equipment—including light bulbs.
6. Electronic and electrical tools.
7. Toys, leisure, and sports equipment.
8. Medical devices (currently exempt)
9. Monitoring and control instruments (currently exempt)
10. Automatic dispensers.
11. Semiconductor devices.

It does not apply to fixed industrial plant and tools. Compliance is the responsibility of the company that puts the product on the market, as defined in the Directive; components and sub-assemblies are not responsible for product compliance. Of course, given the fact that the regulation is applied at the homogeneous material level, data on substance concentrations needs to be transferred through the supply chain to the final producer. An IPC standard has recently been developed and published to facilitate this data exchange, IPC-1752. It is enabled through two PDF forms that are free to use.

RoHS applies to these products in the EU whether made within the EU or imported. Certain exemptions apply, and these are updated on occasion by the EU.

Examples of Product Components Containing Restricted Substances

RoHS restricted substances have been used in a broad array of consumer electronics products. Examples of leaded components include:

- paints and pigments
- PVC (vinyl) cables as a stabilizer (e.g., power cords, USB cables)
- solders
- printed circuit board finishes, leads, internal and external interconnects
- glass in television and photographic products (e.g., CRT television screens and camera lenses)
- metal parts
- lamps and bulbs
- batteries.

Cadmium is found in many of the above components, examples include plastic pigmentation, nickel-cadmium (NiCd) batteries and CdS photocells (used in night lights). Mercury is used in lighting applications and automotive switches, examples include fluorescent lamps (used in laptops for backlighting) and mercury tilt switches (these are rarely used nowadays). Hexavalent chromium is used for metal finishes to prevent corrosion. Polybrominated biphenyls and diphenyl Ethers/Oxides are used primarily as flame retardants.

Hazardous Materials and the High-tech Trash Problem

RoHS and other efforts to reduce hazardous materials in electronics are motivated in part to address the global issue of consumer electronics waste. As newer technology arrives at an ever increasing rate, consumers are discarding their obsolete products sooner than ever. This waste ends up in landfills and in countries like China to be "recycled."

"In the fashion-conscious mobile market, 98 million U.S. cell phones took their last call in 2005. All told, the EPA estimates that in the U.S. that year, between 1.5 and 1.9 million tons of computers, TVs, VCRs, monitors, cell phones, and other equipment were discarded. If all sources of electronic waste are tallied, it could total 50 million tons a year worldwide, according to the UN Environment Programme."

American electronics sent offshore to countries like Ghana in West Africa under the guise of recycling may be doing more harm than good. Not only are adult and child workers in these jobs being poisoned by heavy metals, but these metals are returning to the U.S. "*The U.S. right now is shipping large quantities of leaded materials to China, and China is the world's major manufacturing centre,*" Dr. Jeffrey

Weidenhamer says, a chemistry professor at Ashland University in Ohio. "*It's not all that surprising things are coming full circle and now we're getting contaminated products back.*"

Changing Toxicity Perceptions

In addition to the high-tech trash problem, RoHS reflects contemporary research over the past 50 years in biological toxicology that acknowledges the long-term effects of low-level chemical exposure on populations. New testing is capable of detecting much smaller concentrations of environmental toxins. Researchers are associating these exposures with neurological, developmental, and reproductive changes.

RoHS and other environmental laws are in contrast to historical and contemporary law that seek to address only acute toxicology, that is direct exposure to large amounts of toxins causing severe injury or death.

Life-cycle Impact Assessment of Lead-free Solder

The United States Environmental Protection Agency (EPA) has published a life-cycle assessment (LCA) of the environmental impacts of lead-free and tin-lead solder, as used in electronic products. For bar solders, when only lead-free solders were considered, the tin/copper alternative had the lowest (best) scores. For paste solders, bismuth/tin/silver had the lowest impact scores among the lead-free alternatives in every category except non-renewable resource consumption.

For both paste and bar solders, all of the lead-free solder alternatives had a lower (better) LCA score in toxicity categories than tin/lead solder. This is primarily due to the toxicity of lead, and the amount of lead that leaches from printed wiring board assemblies, as determined by the leachability study conducted by the partnership. The study results are providing the industry with an objective analysis of the life-cycle environmental impacts of leading candidate alternative lead-free solders, allowing industry to consider environmental concerns along with the traditionally evaluated parameters of cost and performance.

This assessment is also allowing industry to redirect efforts toward products and processes that reduce solders' environmental footprint, including energy consumption, releases of toxic chemicals, and potential risks to human health and the environment. Another life-cycle assessment by IKP, University of Stuttgart, shows similar results to those of the EPA study.

Life-cycle Impact Assessment of BFR-free Plastics

The ban on concentrations of brominated flame retardants (BFR) above 0.1% in plastics has had an impact on plastics recycling. As more and more products include recycled plastics, it has become critical to know the BFR concentration in these plastics, either by tracing the origins of the recycled plastics to establish the BFR concentrations, or by measuring the BFR concentrations from samples. Plastics with high BFR concentrations are costly to handle or to discard, whereas plastics with levels below 0.1% have value as recyclable materials.

There are a number of analytical techniques for the rapid measurement of BFR concentrations. X-ray fluorescence spectroscopy can confirm the presence of bromine (Br), but it does not indicate the BFR concentration or specific molecule. Ion attachment mass spectrometry (IAMS) can be used to measure BFR concentrations in plastics. The BFR ban has had significant impacts both upstream—plastic material selection—and downstream—plastic material recycling.

Exemptions

Solar Panels

Cadmium telluride (CdTe) thin-film PV modules in photovoltaic panels are explicitly allowed by RoHS to contain cadmium, even though cadmium is restricted in all other electronics. The solar panel exemption was in the original 2003 RoHS regulation and it was further extended on May 27th, 2011.

Product Category 8 and 9 Exclusions

Medical devices, and monitoring and control instruments comprise RoHS Category 8 and Category 9 products respectively. The EU recognises that these products are manufactured in small numbers and generally have a long product life. Further, these products are often used in mission-critical applications where their failure can reasonably be expected to be extremely disruptive, if not catastrophic. Since the long term effects of lead-free solder, a primary RoHS objective, cannot be known for a period of at least five years following the directive's application to the remaining eight categories, the EU has established at least a temporary moratorium for Category 8 and 9 products.

In an effort to gain more insight the EU commissioned a study to assess when and if the RoHS directive should be applied to Category

8 and 9 products. Released in July 2006, the *Review of Directive 2002/95/EC (RoHS) Categories 8 and 9 – Final Report* recommended that Category 8 and 9 products remain exempt from the RoHS directive until 2012 or 2018 depending upon specific product sub-categories and applications. Since the EU has not yet adopted this recommendation, the exact timing of RoHS application to Category 8 and 9 products remained uncertain.

Legislation published in July, 2011 removes these exemptions.

Labelling

RoHS does not require any specific product labelling, however many manufacturers have adopted their own compliance marks to reduce confusion. Visual indicators in use today include explicit "RoHS compliant" labels, green leaves, check marks, and "PB-Free" markings. In addition, the closely related WEEE (Waste Electrical and Electronic Equipment Directive) trash-can logo with an "X" through it is an indicator that the product may be compliant. Chinese RoHS labels, a lower case "e" within a circle with arrows, can also designate compliance.

The proposed RoHS2 attempts to address this issue by requiring the CE mark, introducing an additional enforcement agency, Trading Standards.

RoHS in Other Regions

Please note that world wide standards and certification are available under the QC 080000 standard, governed by the NSAI (National Standards Authority of Ireland), to ensure the control of RoHs in industrial applications.

Asia / Pacific

China Order No. 39 : Final Measures for the *Administration of the Control and Electronic Information Products* (often referred to as *China RoHS*) has the stated intent to establish similar restrictions, but in fact takes a very different approach. Unlike EU RoHS, where products in specified categories are included unless specifically excluded, there will be a list of included products, known as the *catalogue* — which will be a subset of the total scope of Electronic Information Products, or EIPs, to which the regulations apply. Initially, products that fall under the covered scope must provide markings and disclosure as to the presence of certain substances, while the substances themselves are not (yet) prohibited. There are some products that are EIPs, which are not in scope for EU RoHS, *e.g.* radar systems, semiconductor-manufacturing equipment, photomasks, etc. The list of EIPs is available in Chinese and English. The marking and disclosure aspects of the regulation were intended to take effect on July 1, 2006, but were postponed twice to March 1, 2007. There is no timeline for the catalogue yet.

Japan : Japan does not have any direct legislation dealing with the RoHS substances, but its recycling laws have spurred Japanese manufacturers to move to a lead-free process in accordance with RoHS guidelines. A ministerial ordinance *Japanese industrial standard for Marking Of Specific Chemical Substances* (J-MOSS), effective from July 1, 2006, directs that some electronic products exceeding a specified amount of the nominated toxic substances must carry a warning label.

South Korea : South Korea promulgated the *Act for Resource Recycling of Electrical and Electronic Equipment and Vehicles* on April 2, 2007. This regulation has aspects of RoHS, WEEE, and ELV.

Turkey : Turkey announced the implementation of their Restriction of Hazardous Substances (RoHS) legislation effective June 2009.

North America

California has passed SB 20: Electronic Waste Recycling Act of 2003, or EWRA. This law prohibits the sale of electronic devices after January 1, 2007, that are prohibited from being sold under the EU RoHS directive, but across a much narrower scope that includes LCDs, CRTs, and the like and only covers the four heavy metals restricted by RoHS. EWRA also has a restricted material disclosure requirement.

Effective January 1, 2010, the California Lighting Efficiency and Toxics Reduction Act applies RoHS to general purpose lights, i.e. "lamps, bulbs, tubes, or other electric devices that provide functional illumination for indoor residential, indoor commercial, and outdoor use."

Other US states and cities are debating whether to adopt similar laws, and there are several states that have mercury and PBDE bans already.

Other Standards

RoHS is not the only environmental standard of which electronic product developers should be aware. Manufacturers will find that it is cheaper to have only a single bill of materials for a product that is distributed worldwide, instead of customizing the product to fit each country's specific environmental laws. Therefore, they develop their own standards, which allow only the strictest of all allowable substances.

For example, IBM forces each of their suppliers to complete a Product Content Declaration form to document compliance to their environmental standard Baseline Environmental Requirements for Materials, Parts and Products for IBM Logo Hardware Products. So for example, IBM banned DecaBDE, even though there was formerly a RoHS exemption for this material (overturned by the European Court in 2008).

Similarly, here is Hewlett-Packard's environmental standard: General specification for the environment (GSE).

Criticism

Adverse effects on product quality and reliability, plus high cost of compliance (especially to small business) are cited as criticisms of the directive, as well as early research indicating that the life cycle benefits of lead-free solder versus traditional solder materials are mixed.

Criticism early on came from an industry resistant to change and a misunderstanding of solders and soldering processes. Deliberate misinformation was espoused to resist what was perceived as a "non-tariff barrier created by European bureaucrats." Many believe the industry is stronger now through this experience and has a better understanding of the science and technologies involved.

One criticism of RoHS is that the restriction of lead and cadmium does not address some of their most prolific applications, while being costly for the electronics industry to comply with. Specifically, the total lead used in electronics makes up only 2% of world lead consumption, while 90% of lead is used for batteries (covered by the battery directive, as mentioned above, which requires recycling and limits the use of mercury and cadmium, but does not restrict lead). Another criticism is that less than 4% of lead in landfills is due to electronic components or circuit boards, while approximately 36% is due to leaded glass in monitors and televisions, which can contain up to 2 kg per screen.

The more common Lead-free solder systems have a higher melting point e.g., a 30 °C typical difference for tin-silver-copper alloys but wave soldering temperature is approximately the same at ~255 °C; however at this temperature most typical lead free solders have longer wetting times than eutectic Pb/Sn 37:63 solder. Additionally wetting force is typically lower, which can be disadvantageous (for hole filling), but advantageous in other situations (closely spaced components).

Care must be taken in selection of RoHS solders as some formulations are harder with less ductility, increasing the likelihood of cracks instead of plastic deformation, which is typical for lead-containing solders. Cracks can occur due to thermal or mechanical forces acting on components or the circuit board, the former being more common during manufacturing and the latter in the field. RoHS solders exhibit advantages and disadvantages in these respects, dependent on packaging and formulation.

The editor of Conformity Magazine wonders if the transition to lead-free solder will affect long-term reliability of electronic devices and systems, especially in applications more mission-critical than in consumer products, citing possible breaches due to other environmental factors like oxidation. The refers to the Newark InOne "*RoHS Legislation and Technical Manual*", which cites these and other "lead-free" solder issues, such as:

1. Warping or delamination of printed circuit boards;
2. Damage to through-holes, ICs and components on circuit boards; and,
3. Added moisture sensitivity, all of which may compromise quality and reliability.

Effect on Reliability

Potential reliability concerns are addressed in Annex, item #7, of the RoHS directive, granting servers exemption from regulation until 2010. These issues were raised when the directive was first implemented in 2003 and reliability effects were less known.

Another potential problem that lead-free solders face is the growth of tin whiskers. These thin strands of tin can grow and make contact with an adjacent trace, developing a short circuit. Historically tin whiskers have been associated with a handful of failures, including a nuclear power plant and pacemaker incident where pure tin plating was used. However, these failures pre-date RoHS. They also do not involve consumer electronics, and therefore may employ RoHS-restricted substances if desired. To help mitigate potential problems, lead-free manufacturers are using a variety of approaches such as tin-zinc formulations that produce non-conducting whiskers or formulations that reduce growth, although they do not halt growth completely in all circumstances. Fortunately, experience thus far suggests deployed instances of RoHS compliant products are not failing due to whisker growth. Dr. Ronald Lasky of Dartmouth College reports: "*RoHS has been in force for more than 15 months now, and ~$400B RoHS-compliant products have been produced. With all of these products in the field, no significant numbers of tin whisker-related failures have been reported.*" Whisker growth occurs slowly over time, is unpredictable, and not fully understood, so time may be the only true test of these efforts. Whisker growth is even observable for lead-based solders, albeit on a much smaller scale.

Some countries have exempted medical and telecommunication infrastructure products from the legislation. However, this may be a moot point, as electronic component manufacturers convert their production lines to producing only lead-free parts, conventional parts with eutectic tin-lead solder will simply not be available, even for military, aerospace and industrial users. To the extent that only solder is involved, this is at least partially mitigated by many lead-free components' compatibility with lead-containing solder processes. Leadframe-based components, such as QFPs, SOICs, and SOPs with gull wing leads, are generally compatible since the finish on the part leads contributes a small amount of material to the finished joint. However, components such as BGAs which come with lead-free solder balls and leadless parts are often not compatible with lead-containing processes.

Economic Effect

There are no *de minimus* exemptions, e.g., for micro-businesses. Today, only one micro-business is known to have closed down in order to investigate the cost of compliance. This economic effect was anticipated and at least some attempts at mitigating the effect were made.

Another form of economic effect is the cost of product failures during the switch to RoHS compliance. For example, tin whiskers were responsible for a 5% failure rate in certain components of Swiss Swatch watches in 2006, reportedly triggering a $1 billion recall. Swatch responded to this by applying for exemptions to RoHS compliance for two components. One of these exemptions was effectively approved, with the other still pending after an initial denial. For the denied part Swatch has stated to be using a replacement solder that is almost pure lead, and its application was for permission to switch to a solder with a lower lead content.

Benefits

Health Benefits

RoHS helps reduce damage to people and the environment in third-world countries where much of today's "high-tech trash" ends up. The use of lead-free solders and components has provided immediate health benefits to electronics industry workers in prototype and manufacturing operations. Contact with solder paste no longer represents the same health-hazard it did before.

Reliability Concerns Unfounded

Contrary to the predictions of widespread component failure and reduced reliability, RoHS's first anniversary (July 2007) passed with little fanfare. Today, millions of compliant products are in use worldwide. Most of today's consumer electronics are now RoHS compliant, examples include Apple's iPod portable music players, Dell and HP home computers and servers, Nintendo's Wii, Motorola and Nokia's wireless phones, Netgear routers, and Panasonic televisions and appliances.

Many electronics companies keep "RoHS status" pages on their corporate websites. For example, the AMD website states:

> *"Although lead containing solder cannot be completely eliminated from all applications today, AMD engineers have developed effective technical solutions to reduce lead content in microprocessors and chipsets to ensure RoHS compliance while minimising costs and maintaining product features. There is no change to fit, functional, electrical or performance specifications. Quality and reliability standards for RoHS compliant products are expected to be identical compared to current packages."*

RoHS printed circuit board finishing technologies are surpassing traditional formulations in fabrication thermal shock, solder paste printability, contact resistance, and aluminium wire bonding performance and nearing their performance in other attributes.

The properties of lead-free solder, such as its high temperature resilience, has been used to prevent failures under harsh field conditions. These conditions include 150 °C operating temperatures with test cycles in the range of -40 °C - 150 °C with severe vibration and shock requirements. Automobile manufacturers are turning to RoHS solutions now as electronics move

Flow Properties and Assembly

One of the major differences between lead-containing and lead-free solder pastes is the "*flow*" of the solder in its liquid state. Lead-containing solder has higher surface tension, and tends to move slightly to attach itself to exposed metal surfaces that touch any part of the liquid solder. Lead-free solder conversely tends to stay in place where it is in its liquid state, and attaches itself to exposed metal surfaces only where the liquid solder touches it. This lack of "*flow*"—while typically seen as a disadvantage because it can lead to lesser quality electrical contacts—can be used to place components tighter than they normally could be placed due to the properties of lead-containing solders.

For example, Motorola reports that their new RoHS wireless device assembly techniques are ". . .*enabling a smaller, thinner, lighter unit.*" Their Motorola Q phone would not have been possible without the new solder. The lead-free solder allows for tighter pad spacing.

Some Exempt Products Achieve Compliance

Research into new alloys and technologies is allowing companies to release RoHS products that are currently exempt from compliance, *e.g.* computer servers. IBM has announced a RoHS solution for high lead solder joints once thought to remain a permanent exemption.

The lead-free packaging technology "*...offers economical advantages in relation to traditional bumping processes, such as solder waste reduction, use of bulk alloys, quicker time-to-market for products and a much lower chemical usage rate.*"

Ion-attachment Mass Spectrometry

Ion-attachment mass spectrometry (IAMS) is a form of mass spectrometry that uses a "soft" form of ionization similar to chemical ionization in which a cation is attached to the analyte molecule in a reactive collision:

$$M + X^{+} + A \rightarrow MX^{+} + A$$

Where M is the analyte molecule, X^{+} is the cation and A is a non-reacting collision partner.

Principle

This technique is applicable to gases or any materials that can be vaporized. It uses a non-fragmenting non-conventional ionisation mode, by attachment of a lithium (or alkaline) ion to the gas to be analysed with a more traditional mass filter. This instrument is more dedicated to analysis of moderately-sized molecules such as organic or aromatic compounds.

Applications

Currently, it is used industrially to verify, with a high throughput, the concentrations of brominated flame retardants (BFR) in plastics in compliance with European RoHS (Restriction of Hazardous Substances) regulation in place since 2006. The banned molecules include PBB and PBDE, whose concentration should not exceed 0.1% w/w.

IAMS has also been used to analyze diesel exhaust particles, in ceramic processing and in critical SiO_2 etching during semiconductor manufacturing.

Electronic Waste Recycling Act

The Electronic Waste Recycling Act of 2003 (2003 Cal ALS 526) (EWRA) is a California law to reduce the use of certain hazardous substances in certain electronic products sold in the state. The act was signed into law September 2003.

All CRT, LCD, and plasma display devices contained in televisions, computers, and other electronic equipment with a screen size over 4 inches (10 cm) measured diagonally are covered by the act. After January 1, 2007 these devices may not contain greater than the allowed concentrations of any of these four materials (by weight):

- cadmium : 0.01%
- hexavalent chromium : 0.1%
- lead : 0.1%
- mercury : 0.1%

The Act also requires retailers to collect an Electronic Waste Recycling Fee (effective January 1, 2005) from consumers who purchase covered devices.

Waste Legislation

Waste legislation dictates the way waste should be managed and disposed of by some form of waste management.

China

EU Waste Legislation

- Landfill Directive
- Waste Framework Directive

- Waste Incineration Directive
- WEEE Directive.

UK Waste Legislation

- Animal By-Products Regulations (ABPR)
- Best practicable environmental option (BPEO)
- Certificate of Technical Competence (COTC)
- Control of pollution act
- Environment Act 1995
- Environmental Impact Assessment
- Environmental Protection Act
- Landfill Allowance Trading Scheme (LATS)
- Landfill in the UK
- Landfill tax
- Landfill Tax Regulations
- Waste Management Licensing Regulations
- Statuatory recycling targets.

UK Regulatory Bodies

- Defra
- Environment Agency.

US Legislation

- Comprehensive Environmental Response, Compensation, and Liability Act (CERCLA) "Superfund"
- Medical Waste Tracking Act
- National Environmental Policy Act (NEPA) - Established the Environmental Protection Agency, set out requirements for Environmental Impact Reporting for various kinds of development.
- Resource Conservation and Recovery Act (RCRA) - One of the main pieces of Legislation regarding municipal solid waste, hazardous wastes, and disposal issues.

US Regulatory Bodies

- United States Environmental Protection Agency (EPA) - regulates generation and disposal of hazardous waste
- United States Department of Transportation (DOT) - regulates transportation of hazardous waste

- Nuclear Regulatory Commission (NRC) - regulates nuclear waste.

US State Laws

In addition to laws implementing or advancing portions of the US laws some US states have enacted laws on other waste and environmental subjects.

- California Proposition 65 "The Safe Drinking Water and Toxic Enforcement Act of 1986" - a 1986 California initiative prohibiting the discharge of toxic substances into drinking water sources
- Electronic Waste Recycling Act - a 2003 California law regarding disposal of consumer electronic wastes

Zero Waste

Zero waste is a philosophy that encourages the redesign of resource life cycles so that all products are reused. Any trash sent to landfills and incinerators is minimal. The process recommended is one similar to the way that resources are reused in nature. A working definition of zero waste, often cited by experts in the field originated from a working group of the Zero Waste International Alliance in 2004. The definition is as follows: "Zero Waste is a goal that is ethical, economical, efficient and visionary, to guide people in changing their lifestyles and practices to emulate sustainable natural cycles, where all discarded materials are designed to become resources for others to use. Zero Waste means designing and managing products and processes to systematically avoid and eliminate the volume and toxicity of waste and materials, conserve and recover all resources, and not burn or bury them. Implementing Zero Waste will eliminate all discharges to land, water or air that are a threat to planetary, human, animal or plant health."

In industry this process involves creating commodities out of traditional waste products, essentially making old outputs new inputs for similar or different industrial sectors. An example might be the cycle of a glass milk bottle. The primary input (or resource) is silica-sand, which is formed into glass and then into a bottle. The bottle is filled with milk and distributed to the consumer. At this point, normal waste methods would see the bottle disposed in a landfill or similar. But with a zero-waste method, the bottle can be saddled at the time of sale with a deposit, which is returned to the bearer upon redemption. The bottle is then washed, refilled, and resold. The only material waste is the wash water, and energy loss has been minimised.Zero waste can represent an economical alternative to waste systems, where new resources are continually required to replenish wasted raw materials. It can also represent an environmental alternative to waste since waste represents a significant amount of pollution in the world.

History

1970s: Zero Waste Systems Inc.

The term *zero waste* was first used publicly in the name of a company, Zero Waste Systems Inc (ZWS), which was founded by PhD chemist Paul Palmer in the mid 1970s in Oakland, California. The mission of ZWS was to find new homes for most of the chemicals being excessed by the nascent electronics industry. They soon expanded their services in many other directions.

For example, they accepted free of charge, large quantities of new and usable laboratory chemicals which they resold to experimenters, scientists, companies and tinkerers of every description during the 1970s. ZWS arguably had the largest inventory of laboratory chemicals in all of California, which were sold for half price. They also collected all of the solvent produced by the electronics industry called developer/ rinse (a mixture of xylene and butyl acetate). This was put into small cans and sold as a lacquer thinner. ZWS collected all the "reflow oil" created by the printed circuit industry, which was filtered and resold into the "downhole" (oil well) industry. ZWS pioneered many other projects.

Because they were the only ones in the world in this business, they achieved an international reputation. Many magazine articles were written about them and several television shows featured them. The California Integrated Waste Management Board produced a slide show featuring ZWS's business and the EPA published a number of studies of their business, calling them an "active waste exchange".

The heir to the ZWS mantle is the Zero Waste Institute (ZWI), also founded by Paul Palmer. Building on the lessons learned from ZWS, the ZWI considers recycling to be no more than an appendage to garbage creation and the garbage industry.

ZWI likewise rejects all attempts to reuse garbage or any kind of waste product. Instead, ZWI calls for the redesign of all of the products of industry and commerce, and the processes that produce, sell and make use of them, so that discard never takes place and there is no waste generated needing to be reused or recycled. Discard is seen as the critical step, a commercial and psychological transfer of responsibility which breaks the chain of custody of a product, removes its owner and subjects it to the degradation of garbage management.

The website offers numerous specific examples of ways in which products can be designed so that discard is unnecessary since the lifetime of the product is extended to at least a threshold value of approximately a human lifetime of 100 years. A fully worked out set of principles and analysis is presented, revolving, among other changes, around standardization, modularization and robust design.

A theory of Design Efficiency leading to Design Effectiveness is presented, which means that once a product is designed to be used in perpetuity, it can be fitted out with robust features, strong materials and special conveniences that could not be afforded in a product designed to be discarded after a single use. That theory is applied to packages as an example.

The ZWI rejects all association with the world of recycling, pointing out that there is no theory of recycling in existence; only a trusting hope that it can be useful.

1998-2003: Peak

The movement gained publicity and reached a peak in 1998+2002, and since then has been moving from "theory into action" by focusing on how a "zero waste community" is structured and behaves. The website of the Zero Waste International Alliance has a listing of communities across the globe that have created public policy to promote zero-waste practices. Colorado on a Zero-Waste path and watch a 6-minute video about the zero-waste big picture. Finally, there is a USA zero-waste organisation named the GrassRoots Recycling Network that puts on workshops and conferences about zero-waste activities.

Present Day

The tension between zero waste, viewed as post-discard total recycling of materials only, and zero waste as the reuse of all high level function remains a serious one today. It is probably the defining difference between established recyclers and emerging zero-wasters. A signature example is the difference between smashing a glass bottle (recovering cheap glass) and refilling the bottle (recovering the entire function of the container).

The tension between the literal application of natural processes and the creation of industry-specific more efficient reuse modalities is another tension. Many observers look to nature as an ultimate model for production and innovative materials. Others point out that industrial products are inherently non-natural (such as chemicals and plastics that are mono-molecular) and benefit greatly from industrial methods of reuse, while natural methods requiring degradation and reconstitution are wasteful in that context.

Biodegradable plastic is the most prominent example. One side argues that biodegradation of plastic is wasteful because plastic is expensive and environmentally damaging to make. Whether made of starch or petroleum, the manufacturing process expends all the same materials and energy costs. Factories are built, raw materials are procured, investments are made, machinery is built and used, humans labour and make use of all normal human inputs for education, housing, food etc. Even if the plastic is biodegraded after a single use, all of those costs are lost so it is much more important to design plastic parts for multiple reuse or perpetual lives. The other side argues that keeping plastic out of a dump or the sea is the sole benefit of interest.

Companies moving towards "zero landfill" plants include Subaru, Xerox and Anheuser-Busch.

Recycling

It is important to distinguish recycling from Zero Waste.

Some claim that the key component to zero waste is recycling while others reject that notion in favour of reusing high function. The common understanding of recycling is simply that of placing bottles and cans in a recycle bin. The modern version of recycling is more complicated and involves many more elements of financing and government support. For example, a 2007 report by the U.S. Environmental Protection Agencyý states that the US recycles at a national rate of 33.4% and includes in this figure composted materials. In addition many worldwide commodity industries have been created to handle the materials that are recycled. At the same time, claims of recycling rates have sometimes been exaggerated, for example by the inclusion of soil and organic matter used to cover garbage dumps daily, in the "recycled" column. In states with recycling incentives, there is constant local pressure to pump up the recycling rate figures.

The movement toward recycling has separated itself from the concept of zero waste. One example of this is the computer industry where worldwide millions of PC's are disposed of each year (160 million in 2007). Those computers that enter the recycling stream are broken down into a small amount of raw materials while most merely enter dumps through export to third world countries. Companies are then able to purchase some raw materials, notably steel, copper and glass, reducing the use of new materials. On the other hand, there is an industry, more aligned with the Zero Waste principle of design for long term reuse, that actually repairs computers. It is called the Computer Refurbishing industry and it predates the current campaign to just collect and ship electronics. They have organisations and conferences and have for many years donated computers to schools, clinics and non-profits. Zero Waste planning demands that components be redesigned for effective reuse over long lives leading to even more refurbishing and repair.

There is one seminal example that brings out the difference between Zero Waste and recycling in stark relief. That example, quoted in Getting To Zero Waste, is the software business. Zero Waste is sensitive to the waste of intellectual effort that would be caused by the need to recreate certain basic inventions of software (called objects in software design) as opposed to copying them over and over whenever needed.

The waste would occur as the software developers consume resources while solving problems already solved earlier. The application of Zero Waste analysis is straightforward as it recommends conserving human effort. On the other hand, the usual approach of recycling would be to look for some materials that could be found to reuse. The materials on which software is saved (such as paper or diskettes)is of little significance compared to the saving of human effort and if software is saved electronically, there is no media at all. Thus Zero Waste correctly identifies a wasteful behaviour to avoid while recycling has no application.

The recycling movement has been embraced by the garbage industry because it serves so well as greenwashing i.e. a way to show that design for garbage creation is acceptable because materials will be kept out of a dump by recycling them. Zero Waste, on the other hand, offers the garbage industry no such screen against public condemnation of waste, and therefore actually threatens the continued need for garbage disposal. For example, in Alameda County, California, garbage dumping is charged a surcharge of $8/ton (as of 2009) which goes entirely for a recycling subsidy but none of which goes for any kind of Zero Waste style designing. Zero Waste has received no support from the garbage industry or politicians under their control except in those cases where it can be claimed to consist solely of more recycling.

Reduce and Reuse

Zero waste is poorly supported by the enactment of government laws to enforce the waste hierarchy of reduce, reuse, and recycle. In practice, these laws invariably emphasize destruction and recycling, while the reuse component is marginalized.

A special feature of Zero Waste as a design principle is that it can be applied to any product or process, in any situation or at any level. Thus it applies equally to toxic chemicals as to benign plant matter. It applies to the waste of atmospheric purity by coal burning or the waste of radioactive resources by attempting to designate the excesses of nuclear power plants as “nuclear waste”. All processes can be designed to minimise the need for discard, both in their own operations and in the usage or consumption patterns which the design of their products leads to. Recycling, on the other hand, deals only with simple materials.

Zero Waste can even be applied to the waste of human potential by enforced poverty and the denial of educational opportunity. It encompasses redesign for reduced energy wasting in industry or transportation and the wasting of the earth’s rainforests. It is a general principle of designing for the efficient use of all resources, however defined.

The recycling movement may be slowly branching out from its solid waste management base to include issues that are similar to the community sustainability movement.

Zero waste on the other hand, is not based in waste management limitations to begin with but requires that we maximise our existing reuse efforts while creating and applying new methods that minimise and eliminate destructive methods like incineration and recycling. Zero Waste strives to ensure that products are designed to be repaired, refurbished, remanufactured and generally reused.. (“What is Zero Waste?”).

The Significance of Dump Capacity

Many dumps are currently exceeding carrying capacity. This is often, mistakenly used as a justification for moving to Zero Waste. Others counter by pointing out that there are huge tracts of land available throughout the USA and other countries which could be used for dumps. This is no more of an argument against the need for Zero Waste than is the former an argument for Zero Waste. The underlying need to move to a society designed along Zero Waste principles arises from the huge waste of resources that is inherent in poorly made, short-lived articles and production processes. The locus of the most egregious wasting takes place as articles are built and processes are run wastefully.

The actual placing of a now useless item in a dump is barely the icing on the cake, in terms of the waste it represents. Poorly conceived proposals, that appear with a dismaying regularity on the Internet, to blithely destroy all garbage as a way to solve the garbage problem, make use of the common delusion that it is the garbage itself which is the problem. These proposals typically claim to convert all or a large portion of existing garbage into oil and sometimes claim to produce so much oil that the world will henceforth have abundant liquid fuels. One such plan, called Anything Into Oil was promoted by Discover Magazine and Fortune Magazine in 2004, even though it absurdly claimed to be able to convert a refrigerator into "light Texas crude" by the application of high pressure steam. Zero Waste analysis, which is long on scientific results and short on spectacular claims, receives no such promotion by the media.

Corporate Initiatives

An example of a company that has demonstrated a change in landfill waste policy is General Motors (GM). GM has confirmed their plans to make approximately half of its 181 plants worldwide "landfill-free" by the end of 2010. Companies like Subaru, Toyota, and Xerox are also producing landfill-free plants. GM is supposed to have about eighty producing plants twenty months. Furthermore, The United States Environmental Protection Agency (EPA) has worked with GM and other companies for decades to minimise the waste through its WasteWise program.

The goal for General Motors is finding ways to recycle or reuse more than 90% of materials by: selling scrap materials, adopting reusable parts boxes to replace cardboard, and even recycling used work gloves. The remainder of the scraps might be incinerated to create energy for the plants. Besides being nature friendly, it also saves money by cutting out waste and producing a more efficient production. All these organisations push forth to make our world clean and producing zero waste.

Re-Use of Waste

The waste sent to landfills may be harvested as useful materials, such as in the production of solar energy or fertiliser for crops.

It may also be reused and recycled for something that we can actually use. "The success of General Motors in creating zero-landfill facilities shows that zero-waste goals can be a powerful impetus for manufacturers to reduce their waste and carbon footprint," says Latisha Petteway, a spokesperson for the EPA.

Construction and Deconstruction

Zero Waste is a goal, a process, a way of thinking that profoundly changes our approach to resources and production. Zero Waste is not about recycling and diversion from landfills but about restructuring production and distribution systems to prevent waste from being manufactured in the first place.

The materials that are still required in these re-designed, resource-efficient systems will be reused many times as the products that incorporate them are reused. Deconstruction can be described as construction in reverse. It involves carefully taking apart a building to maximise the reuse of materials, thereby reducing waste and conserving resources. Deconstruction can capture materials and some components from the millions of buildings that are existing and that were poorly designed for high level reuse but it is not a favoured approach from a Zero Waste point of view. Zero Waste favours the design of buildings as assemblages of high level components, not their creation from rough materials such as lumber, cement or plaster.

The details are not worked out yet but to the extent that entire rooms, entire walls, roofs or floors or entire utility systems can be pre-built and installed as completed components, that will be the goal of Zero Waste design. Until buildings are built as components capable of later dismantling, deconstruction is a stop-gap process that the United States can use to minimise the waste of building materials. For now, the largest parts that we are able to save tend to be architectural elements, windows, doors, and metals, many of which are being saved and resold by reuse yards such as Urban Ore in Berkeley California.

The main parts that still need to be crushed are wood flooring, brick walls, and structural timbers. The demolition of traditional buildings has been long done by wrecking ball or bulldozer. Social and political artifacts, such as demolition contractor licenses and required permits that can only be satisfied by destruction and discard (with partial recycling of rubble and steel), render the destruction and disposal costs cheaper than deconstruction. Approximately seventy pounds of the waste is generated for about every square foot of the residential building demolition. It is arguable that this is artificial economics, based on the cultural preference for wastefulness and that

Zero Waste designs of dismantlable components will ultimately be the cheapest as well as the most conservative way to reuse buildings. Further discussions of this topic may be found on the ZWI website.

Market-based Campaigns

Market-based, legislation-mediated campaigns like Extended Producer Responsibility (EPR) and the Precautionary Principle are among numerous campaigns that have a Zero Waste slogan hung on them by means of claims they will ineluctably lead to policies of Zero Waste. At the moment, there is no evidence that EPR will increase reuse, rather than merely moving discard and disposal into private-sector dumping contracts.

The Precautionary Principle is put forward to shift liability for proving new chemicals are safe from the public (acting as guinea pig) to the company introducing them. As such, its relation to Zero Waste is dubious. Likewise, many organisations, cities and counties have embraced a Zero Waste slogan while pressing for none of the key Zero Waste changes. In fact, it is common for many such to simply state that recycling is their entire goal.

Many commercial or industrial companies claim to embrace Zero Waste but usually mean no more than a major materials recycling effort, having no bearing on product redesign. Examples include Staples, Home Depot, Toyota, General Motors and computer take-back campaigns. Earlier social justice campaigns have successfully pressured McDonald's to change their meat purchasing practices and Nike to change its labour practices in Southeast Asia. Those were both based on the idea that organised consumers can be active participants in the economy and not just passive subjects. However, the announced and enforced goal of the public campaign is critical. A goal to reduce waste generation or dumping through greater recycling will not achieve a goal of product redesign and so cannot reasonably be called a Zero Waste campaign.

Governance

Re-shaping people's resource use pattern is a challenge beyond the scope of piecemeal and reactive laws. Policy incrementalism reflects policymakers tendency to build on what already exists and is characterised by minimal disturbance of the current state. This cannot achieve the targets of Zero Waste at a national level. What is needed is a shift from government (formal organisations and procedures of the public sector) to governance (array of governmental and non-governmental institutions). The government alone does not have the cognitive breadth to determine how to reach the ultimate target of Zero Waste which requires the involvement of businesses, NGOs, the public and the state in governing. Governance is "ultimately concerned with creating the conditions for ordered rule and collective action" and to achieve Zero Waste, a departure from waste management based simply on waste disposal ideology is necessary.

The role of the state is to act as central steering mechanism and governance networks that bring together government; public and market actors are viewed as important to achieve Zero Waste similar to many other environmental goals. The Johannesburg World Summit on Sustainable Development emphasized these partnerships between different stakeholders in the environment and the establishment of Public-Private partnerships is seen as the best way to achieve sustainability.

An example of network governance approach can be seen in the UK under New Labour who proposed the establishment of regional groupings that brought together the key stakeholders in waste management (local authority representatives, waste industry, government offices etc.) on a voluntary basis. There is a lack of clear government policy on how to meet the targets for diversion from landfill which increases the scope at the regional and local level for governance networks.

The overall goal is set by government but the route for how to achieve it is left open. Governance in waste management seeks to widen the range of stakeholders involved and improve co-ordination between them. This mobilises a collective action which is essential to overcome potential conflicts when tackling a goal as visionary as Zero Waste.

The challenge of governance in waste management therefore is how to get collective action across the broad spectrum of stakeholders. Zero Waste is a strategy promoted by environmental NGOs but the waste industry is more in favour of the capital intensive option of energy from waste incineration. Research often highlights public support as the first requirement for success. In Taiwan, public opinion was essential in changing the attitude of business, who must transform their material use pattern to become more sustainable for Zero Waste to work.

The public were made aware of the importance of sustainability through communication with governmental and nongovernmental organisations illustrating the importance of networks. The latest development in Zero Waste is the city of Masdar in Abu Dhabi which promises to be a Zero Waste city. Innovation and technology is encouraged by government creating an innovation friendly environment without being prescriptive. To be a successful model of sustainable urban development it will also require the involvement and co-operation from all members of society emphasizing the importance of network governance.

Plastic Recycling

Plastic recycling is the process of recovering scrap or waste plastics and reprocessing the material into useful products, sometimes completely different in form from their original state. For instance, this could mean melting down soft drink bottles and then casting them as plastic chairs and tables. Typically a plastic is not recycled into the same type of plastic, and products made from recycled plastics are often not recyclable.

Challenges

When compared to other materials like glass and metal materials, plastic polymers require greater processing (Heat treating, Thermal depolymerization and monomer recycling) to be recycled. Plastics have a low entropy of mixing, which is due to the high molecular weight of their large polymer chains. A macromolecule interacts with its environment along its entire length, so its enthalpy of mixing is large compared to that of an organic molecule with a similar structure. Heating alone is not enough to dissolve such a large molecule; because of this, plastics must often be of nearly identical composition in order to mix efficiently.

When different types of plastics are melted together they tend to phase-separate, like oil and water, and set in these layers. The phase boundaries cause structural weakness in the resulting material, meaning that polymer blends are only useful in limited applications.

Another barrier to recycling is the widespread use of dyes, fillers, and other additives in plastics. The polymer is generally too viscous to economically remove fillers, and would be damaged by many of the processes that could cheaply remove the added dyes. Additives are less widely used in beverage containers and plastic bags, allowing them to be recycled more often. Yet another barrier to removing large quantities of plastic from the waste stream and landfills is the fact that many common but small plastic items lack the universal triangle recycling symbol and accompanying number. A perfect example is the billions of plastic utensils commonly distributed at fast food restaurants or sold for use at picnics.

The use of biodegradable plastics is increasing. If some of these get mixed in the other plastics for recycling, the reclaimed plastic is not recyclable because of the variance in properties and melt temperatures.

Processes

Before recycling, most plastics are sorted according to their resin type. In the past, plastic reclaimers used the resin identification code (RIC), a method of categorization of polymer types, which was developed by the Society of the Plastics Industry in 1988. Polyethylene terephthalate, commonly referred to as PET, for instance, has a resin code of 1. Most plastic reclaimers do not rely on the RIC now; they use automatic sort systems to identify the resin, such as near infrared (NIR) technology. Some plastic products are also separated by colour before they are recycled. The plastic recyclables are then shredded. These shredded fragments then undergo processes to eliminate impurities like paper labels. This material is melted and often extruded into the form of pellets which are then used to manufacture other products.

Monomer Recycling

Many recycling challenges can be resolved by using a more elaborate *monomer recycling* process, in which a condensation polymer essentially undergoes the inverse of the polymerization reaction used to manufacture it. This yields the same mix of chemicals that formed the original polymer, which can be purified and used to synthesize new polymer chains of the same type. Du Pont opened a pilot plant of this type in Cape Fear, North Carolina, USA, to recycle PET by a process of methanolysis, but it closed the plant due to economic pressures.

Thermal Depolymerization

Another process involves the conversion of assorted polymers into petroleum by a much less precise thermal depolymerization process. Such a process would be able to accept almost any polymer or mix of polymers, including thermoset materials such as vulcanized rubber tires and the biopolymers in feathers and other agricultural waste. Like natural petroleum, the chemicals produced can be made into fuels as well as polymers. A pilot plant of this type exists in Carthage, Missouri, USA, using turkey waste as input material. Gasification is a similar process, but is not technically recycling since polymers are not likely to become the result.

Heat Compression

Yet another process that is gaining ground with startup companies (especially in Australia, United States and Japan) is heat compression. The heat compression process takes all unsorted, cleaned plastic in all forms, from soft plastic bags to hard industrial waste, and mixes the load in tumblers (large rotating drums resembling giant clothes dryers). The most obvious benefit to this method is the fact that all plastic is recyclable, not just matching forms. However, criticism rises from the energy costs of rotating the drums, and heating the post-melt pipes?

Other Processes

A process has also been developed in which many kinds of plastic can be used as a carbon source in the recycling of scrap steel.

Applications

PET

Post-consumer polyethylene terephthalate (PET) containers are sorted into different colour fractions, and baled for onward sale. PET recyclers further sort the baled bottles and they are washed and flaked (or flaked and then washed). Non PET fractions such as caps and labels are removed during this process. The clean flake is dried. Further treatment can take place e.g. melt filtering and pelletising or various treatments to produce food contact approved recycled PET (RPET).

RPET has been widely used to produce polyester fibres. This sorted post-consumer PET waste is crushed, chopped into flakes, pressed into bales, and offered for sale.

One use for this recycled PET that has recently started to become popular is to create fabrics to be used in the clothing industry. The fabrics are created by spinning the PET flakes into thread and yarn. This is done just as easily as creating polyester from brand new PET. The recycled PET thread or yarn can be used either alone or together with other fibres to create a very wide variety of fabrics. Traditionally these fabrics were used to create strong, durable, rough, products, such as jackets, coat, shoes, bags, hats, and accessories. However, these fabrics are usually too rough on the skin and could cause irritation. Therefore, they usually are not used on any clothing that may irritate the skin, or where comfort is required. But in today's new eco-friendly world there has been more of a demand for "green" products. As a result, many clothing companies have started looking for ways to take advantage of this new market and new innovations in the use of recycled PET fabric are beginning to develop. These innovations included different ways to process the fabric, to use the fabric, or blend the fabric with other materials.

Some of the fabrics that are leading the industry in these innovations include Billabong's Eco-Supreme Suede, Livity's Rip-Tide III, Wellman Inc's Eco-fi(formerly known as EcoSpun), and Reware's Rewoven. Some additional companies that take pride in using recycled PET in their products are Crazy Shirts and Playback.

Other major outlets for RPET are new containers (food contact or non food contact) produced either by (injection stretch blow) moulding into bottles and jars or by thermoforming APET sheet to produce clam shells, blister packs and collation trays. These applications used 46% of all RPET produced in Europe in 2010. Other applications, such as strapping tape, injection moulded engineering components and even building materials account for 13% of the 2010 RPET production.

PVC

In Europe, developments in PVC waste management are monitored by Vinyl 2010, a legal entity established in 2000. In the waste management area their commitment is to

1. Support integrated waste management approaches, using raw materials as efficiently as possible
2. Work with the stakeholders to research, develop, and implement recycling of 200,000 tonnes per year of PVC postconsumer waste in 2010 in addition to waste already recycled in 2000, or regulated by the PPW, ELV and E&E Waste Directives
3. Recycle collectable, available PVC postconsumer waste from pipes, window profiles, and roofing membranes.

Vinyl 2010 has a Monitoring Committee and publishes annual reviews. In 2011, it reported that 260,842 tonnes of post-consumer PVC waste was recycled in 2010, i.e. an increase of 220,000 tonnes over the 1999 volumes, exceeding the 10-year target of 200,000 tonnes.

Collection and recycling schemes for PVC waste stream are managed through Recovinyl which reported the recycled tonnage as follows: profile 107,000 tonnes, flexible cables 79,000 tonnes, pipe 25,000 tonnes, rigid film 6,000 tonnes, and mixed flexible 38,000 tonnes. Recovinyl states that of the recycled material 75% is for floors, 15% for foils, 5% for traffic cones, 3% for hoses end 2% for other applications. One of the recycling processes is the Vinyloop Texyloop used for solvent-based mechanical recycling. It involves recovering PVC plastic from composite materials through dissolution and precipitation, and is a closed-loop system, recycling the solvent and regenerating PVC.

HDPE

The most-often recycled plastic HDPE (high-density polyethylene) or number 2, is downcycled into plastic lumber, tables, roadside curbs, benches, truck cargo liners, trash receptacles, stationery (e.g. rulers) and other durable plastic products and is usually in demand.

PS

Most polystyrene products are currently not recycled due to the lack of incentive to invest in the compactors and logistical systems required. Expanded polystyrene scrap can be easily added to products such as EPS insulation sheets and other EPS materials for construction applications. And many manufacturers cannot obtain sufficient scrap because of the aforementioned collection issues. When it is not used to make more EPS, foam scrap can be turned into clothes hangers, park benches, flower pots, toys, rulers, stapler bodies, seedling containers, picture frames, and architectural molding from recycled PS. Recycled EPS is also used in many metal casting operations. Rastra is made from EPS that is combined with cement to be used as an insulating amendment in the making of concrete foundations and walls. American manufacturers have produced insulating concrete forms made with approximately 80% recycled EPS since 1993.

Other Plastics

The white plastic polystyrene foam peanuts used as packing material are often accepted by shipping stores for reuse.

Successful trials in Israel have shown that plastic films recovered from mixed municipal waste streams can be recycled into useful household products such as buckets.

Similarly, agricultural plastics such as mulch film, drip tape and silage bags are being diverted from the waste stream and successfully recycled into much larger products for industrial applications such as plastic composite railroad ties. Historically, these agricultural plastics have primarily been either landfilled or burned on-site in the fields of individual farms.

CNN reports that Dr. S. Madhu of the Kerala Highway Research Institute, India has formulated a road surface that includes recycled plastic. Aggregate, bitumen (asphalt) with plastic that has been shredded and melted at a temperature below 220 degrees C (428 °F) to avoid pollution. This road surface is claimed to be very durable and monsoon rain resistant. The plastic is sorted by hand, which is economical in India.

The test road used 60 kg of plastic for an approx. 500m long, 8m wide, two-lane road. The process chops thin-film road-waste into a light fluff of tiny flakes that hot-mix plants can uniformly introduce into viscous bitumen with a customized dosing machine. Tests at both Bangalore and the Indian Road Research Centre indicate that roads built using this 'KK process' will have longer useful lives and better resistance to cold, heat, cracking, and rutting, by a factor of three.

Recycling Rates

The quantity of post-consumer plastics recycled has increased every year since at least 1990, but rates lag far behind those of other items, such as newspaper (about 80%) and corrugated fibreboard (about 70%). Overall U.S. post-consumer plastic waste for 2008 was estimated at 33.6 million tons; 2.2 million tons (6.5%) were recycled and 2.6 million tons (7.7%) were burned for energy; 28.9 million tons, or 85.5%, were discarded in landfills.

Economic and Energy Potential

In 2008, the price of PET dropped from $370/ton in the US to $20 in November. PET prices had returned to their long term averages by May 2009.

Consumer Education

United States

Low national plastic recycling rates have been due to the complexity of sorting and processing, unfavourable economics, and consumer confusion about which plastics can actually be recycled. Part of the confusion has been due to the recycling symbol that is usually on all plastic items. This symbol is called a resin identification code. It is stamped or printed on the bottom of containers and surrounded by a triangle of arrows. The intent of these arrows was to make it easier to identify plastics for recycling. The recycling symbol doesn't necessarily mean that the item will be accepted by residential recycling programs.

United Kingdom

In the UK, the amount of post-consumer plastic being recycled is relatively low, due in part to a lack of recycling facilities.

The Plastics 2020 Challenge was founded in 2009 by the plastics industry with the aim of engaging the British public in a nationwide debate about the use, reuse and disposal of plastics, hosts a series of online debates on its website framed around the waste hierarchy.

Plastic Identification Code

Five groups of plastic polymers, each with specific properties, are used worldwide for packaging applications. Each group of plastic polymer can be identified by its Plastic Identification code (PIC) - usually a number or a letter abbreviation. For instance, Low-Density Polyethylene can be identified by the number "4" or the letters "LDPE". The PIC appears inside a three-chasing arrow recycling symbol. The symbol is used to indicate whether the plastic can be recycled into new products.

The PIC was introduced by the Society of the Plastics Industry, Inc., which provides a uniform system for the identification of different polymer types and helps recycling companies to separate different plastics for reprocessing. Manufacturers of plastic products are required to use PIC labels in some countries/regions and can voluntarily mark their products with the PIC where there are no requirements. Consumers can identify the plastic types based on the codes usually found at the base or at the side of the plastic products, including food/ chemical packaging and containers. The PIC is usually not present on packaging films, as it is not practical to collect and recycle most of this type of waste.

Plastics 2020 Challenge

The Plastics 2020 Challenge was founded on 7 July 2009 with the aim of engaging the public in a nationwide debate about the use, reuse and disposal of plastics.

As part of the UK plastics industry's pledge to lead the UK in diverting plastics from landfill by 2020 and to inform the public about the environmental challenges facing society today, the Challenge hosts a series of online debates on its website framed around the Four R's: reduce, reuse, recycle, recover.

The Plastics 2020 Challenge was named runner-up "Campaign of the Year" for 2009 by *Packaging News*.

Sponsors

The Plastics 2020 Challenge is sponsored by three organisations:

- Plastics Europe
- British Plastics Federation (BPF)
- Packaging and Films Association (PAFA).

Microplastics

Microplastics are small plastic particles in the environment and have become a paramount issue especially in the marine environment. Not unequivocally defined, some marine researchers define microplastics as all plastic particles smaller than 1 mm pertaining to their microscopic size range while others in turn define them as smaller than 5 mm recognising the common use of 333 ìm mesh neuston nets for field sampling. However, their integral impact on wildlife is scientifically not well established yet.

There are two main sources of microplastics:

1. Microplastics which are produced either for direct use, such as for industrial abrasives, exfoliants, cosmetics or rotomilling or for indirect use as precursors (so called resin pellets or nurdles) for the production of manifold consumer products ("primary microplastics").
2. Microplastics formed in the environment as a consequence of the breakdown of larger plastic material, especially marine debris, into smaller and smaller fragments (so called "secondary microplastics"). The breakdown is caused by mechanical forces (e.g. waves) and/or photochemical processes triggered by sunlight (especially UVB).

The abundance and global distribution of microplastics in the oceans has steadily increased over the last few decades with rising plastic consumption worldwide.

Potential Impacts on the Marine Environment

The scientists that participated in the first International Research Workshop on the Occurrence, Effects and Fate of Microplastic Marine Debris, held September 9–11, 2008 on the University of Washington Tacoma campus in Tacoma, Washington, USA, agreed that microplastics may pose problems in the marine environment based on the following:

1. the documented occurrence of microplastics in the marine environment,

2. the long residence times of these particles (and, therefore, their likely buildup in the future), and
3. their demonstrated ingestion by marine organisms.

So far, research has mainly focused on larger plastic items. Widely recognised problems are associated with entanglement, ingestion, suffocation and general debilitation often leading to death and/or strandings. This raises serious public concern. In contrast, microplastics are not as conspicuous, being less than 5 mm. Particles of this size are available to a much broader range of species and have been shown to be ingested by deposit-feeding lugworms (Arenicola marina) and filter-feeding mussels (Mytilus edulis) to name just two examples.

Ingestion of microplastics by species at the base of the food web causes concern as little is known about its effects. It remains unknown if microplastics may be transferred across trophic levels.

Possible effects of microplastics on marine organisms after ingestion are threefold:

1. physical blockage or damage of feeding appendages or digestive tract,
2. leaching of plastic component chemicals into organisms after digestion, and
3. ingestion and accumulation of sorbed chemicals by the organism."

Small animals are at risk of reduced food intake due to false satiation and resulting starvation or other physical harm. However, long term impacts on marine organisms are currently unknown. Plastic debris has also been shown to serve as carrier for the dispersal of biota, thus greatly increasing dispersal opportunities in the oceans, endangering marine biodiversity worldwide. The dispersal of aggressive alien and invasive species is as much a topic as the dispersal of cosmopolitan species. Approximately half of the plastic material introduced to the marine environment is buoyant, but fouling by organisms can induce the sinking of additional plastic debris to the sea floor, where it may interfere with sediment-dwelling species and sedimental gas exchange processes. However, this is of more importance for larger plastic debris.

Microplastics and Persistent Organic Pollutants (POPs)

Furthermore, plastic particles may highly concentrate and transport synthetic organic compounds (e.g. persistent organic pollutants, POPs) commonly present in the environment and ambient sea water on their surface through adsorption. It still remains unknown, if microplastics can act as agents for the transfer of POPs from the environment to organisms in this way but evidence suggest this to be a potential portal for entering food webs. Not only do POPs raise concern. Additives added to plastics during manufacture may leach out upon ingestion, potentially causing serious harm to the organism. Endocrine disruption by plastic additives may affect the reproductive health of humans and wildlife alike. At current levels, microplastics are unlikely to be an important global geochemical reservoir for POPs such as PCBs, dioxins, and DDT in open oceans. It is not clear, however, if microplastics play a larger role as chemical reservoirs on smaller scales. A reservoir function is conceivable in densely populated and polluted areas, such as bights of mega-cities, areas of intensive agriculture and effluents flumes.

Oil based polymers ('plastics') are virtually non-biodegradable. However, renewable natural polymers are now in development which can be used for the production of biodegradable materials similar to that of oil-based polymers. Their properties in the environment, however, require detailed scrutiny before their wide use is propagated.

Lumber

Lumber or timber is wood in any of its stages from felling through readiness for use as structural material for construction, or wood pulp for paper production. (The distinction between the two terms is discussed below.) Lumber is supplied either rough or finished. Besides pulpwood, *rough lumber* is the raw material for furniture-making and other items requiring additional cutting and shaping. It is available in many species, usually hardwoods. *Finished lumber* is supplied in standard sizes, mostly for the construction industry, primarily softwood from coniferous species including pine, fir and spruce (collectively known as Spruce-pine-fir), cedar, and hemlock, but also some hardwood, for high-grade flooring.

Terminology

In the United Kingdom and other Commonwealth Countries such as Australia and New Zealand, *timber* is a term also used for sawn wood products, for example timber floor boards, where as generally in the United States and Canada, the product of timber cut into boards is referred to as *lumber*. In the United States and Canada, timber often refers to the wood contents of standing, live trees that can be used for lumber or fibre production, although it can also be used to describe sawn lumber whose smallest dimension is not less than 5 inches (127 mm) such as the large dimension and often partially finished lumber used in timber-frame construction. In the United Kingdom the word lumber has several other meanings, including unused or unwanted items. Note that the word lumberjack is used in the UK and Australia to refer to North Americans who fell standing trees, and so the word *lumber* conjures images of what North Americans call *timber*, and vice versa. "Timber!" is also an exclamation that lumberjacks often shout out to warn others that a cut tree is about to fall.

Dimensional Lumber

Dimensional lumber is a term used for lumber that is finished/ planed and cut to standardized width and depth specified in inches. Examples of common sizes are *2×4* (pictured, also *two-by-four* and other variants, such as *four-by-two* in the UK, Australia, New Zealand), *2×6*, and *4×4*. The length of a board is usually specified separately from the width and depth. It is thus possible to find 2×4s that are four, eight, or twelve feet in length. In the United States and Canada the standard lengths of lumber are 6, 8, 10, 12, 14, 16, 18, 20, 22, and 24 feet. For wall framing, "stud," or "precut" sizes are available, and commonly used. For an eight, nine, or ten foot ceiling height, studs are available in 92 5/8 inches, 104 5/8 inches, and 116 5/8 inches.

Table : North American softwood dimensional lumber sizes

Nominal (in)	*Actual*	*Nominal (in)*	*Actual*	*Nominal (in)*	*Actual*
1 × 2	¾ in × 1 ½ in (19 mm × 38 mm)	2 × 2	1 ½ in × 1 ½ in (38 mm × 38 mm)	4 × 4	3 ½ in × 3 ½ in (89 mm × 89 mm)
1 × 3	¾ in × 2 ½ in (19 mm × 64 mm)	2 × 3	1 ½ in × 2 ½ in (38 mm × 64 mm)	4 × 6	3 ½ in × 5 ½ in (89 mm × 140 mm)
1 × 4	¾ in × 3 ½ in (19 mm × 89 mm)	2 × 4	1 ½ in × 3 ½ in (38 mm × 89 mm)	6 × 6	5 ½ in × 5 ½ in (140 mm × 140 mm)
1 × 6	¾ in × 5 ½ in (19 mm × 140 mm)	2 × 6	1 ½ in × 5 ½ in (38 mm × 140 mm)	8 × 8	7 ¼ in × 7 ¼ in (184 mm × 184 mm)
1 × 8	¾ in × 7 ¼ in (19 mm × 184 mm)	2 × 8	1 ½ in × 7 ¼ in (38 mm × 184 mm)		
1 × 10	¾ in × 9 ¼ in (19 mm × 235 mm)	2 × 10	1 ½ in × 9 ¼ in (38 mm × 235 mm)		
1 × 12	¾ in × 11 ¼ in (19 mm × 286 mm)	2 × 12	1 ½ in × 11 ¼ in (38 mm × 286 mm)		

Solid dimensional lumber typically is only available up to lengths of 24 ft. Engineered wood products, manufactured by binding the strands, particles, fibres, or veneers of wood, together with adhesives, to form composite materials, offer more flexibility and greater structural strength than typical wood building materials.

Pre-cut studs save a framer a lot of time as they are pre-cut by the manufacturer to be used in 8 ft, 9 ft & 10 ft ceiling applications, which means they have removed a few inches of the piece to allow for the sill plate and the double top plate with no additional sizing necessary.

In the Americas, *two-bys* (2×4s, 2×6s, 2×8s, 2×10s, and 2×12s), along with the 4×4, are common lumber sizes used in modern construction. They are the basic building block for such common structures as balloon-frame or platform-frame housing. Dimensional lumber made from softwood is typically used for construction, while hardwood boards are more commonly used for making cabinets or furniture.

Lumber's *nominal* dimensions are given in terms of green (not dried), rough (unfinished) dimensions. The *finished* size is smaller, as a result of drying (which shrinks the wood), and planing to smooth the wood. However, the difference between "nominal" and "finished" lumber size can vary. So various standards have specified the difference between nominal size, and finished size, of lumber.

Early standards called for green rough lumber to be of full nominal dimension when dry, but the requirements have changed over time. For example, in 1910, a typical finished 1-inch- (25 mm) board was $^{13}/_{16}$ in (21 mm). In 1928, that was reduced by 4%, and yet again by 4% in 1956. In 1961, at a meeting in Scottsdale, Arizona, the Committee on Grade Simplification and Standardization agreed to what is now the current U.S. standard: in part, the dressed size of a 1 inch (nominal) board was fixed at $^{3}/_{4}$ inch; while the dressed size of 2 inch (nominal) lumber was *reduced* from 1 $^{5}/_{8}$ inch to the now standard 1 $^{1}/_{2}$ inch.

Grades and Standards

Individual pieces of lumber exhibit a wide range in quality and appearance with respect to knots, slope of grain, shakes and other natural characteristics. Therefore, they vary considerably in strength, utility and value.

The move to set national standards for lumber in the United States began with publication of the American Lumber Standard in 1924, which set specifications for lumber dimensions, grade, and moisture content; it also developed inspection and accreditation programs. These standards have changed over the years to meet the changing needs of manufacturers and distributors, with the goal of keeping lumber competitive with other construction products. Current standards are set by the American Lumber Standard Committee, appointed by the Secretary of Commerce.

Design values for most species and grades of visually graded structural products are determined in accordance with ASTM standards, which consider the effect of strength reducing characteristics, load duration, safety and other influencing factors. The applicable standards are based on results of tests conducted in cooperation with the USDA Forest Products Laboratory. Design Values for Wood Construction, which is a supplement to the ANSI/AF&PA National Design Specification® for Wood Construction, provides these lumber design values, which are recognised by the model building codes. A summary of the six published design values—including bending (Fb), shear parallel to grain (Fv), compression perpendicular to grain (Fc-perp), compression parallel to grain (Fc), tension parallel to grain (Ft), and modulus of elasticity (E and Emin) can be found in Structural Properties and Performance published by WoodWorks.

Canada has grading rules that maintain a standard among mills manufacturing similar woods to assure customers of uniform quality. Grades standardize the quality of lumber at different levels and are based on moisture content, size and manufacture at the time of grading, shipping and unloading by the buyer. The National Lumber Grades Authority (NLGA) is responsible for writing, interpreting and maintaining Canadian lumber grading rules and standards. The Canadian Lumber Standards Accreditation Board (CLSAB) monitors the quality of Canada's lumber grading and identification system.

Attempts to maintain lumber quality over time have been challenged by historical changes in the timber resources of the United States—from the slow-growing virgin forests common over a century ago to the fast-growing plantations now common in today's commercial forests. Resulting declines in lumber quality have been of concern to both the lumber industry and consumers and have caused increased use of alternative construction products

Machine stress-rated and machine-evaluated lumber is readily available for end-uses where high strength is critical, such as truss rafters, laminating stock, I-beams and web joints. Machine grading measures a characteristic such as stiffness or density that correlates with the structural properties of interest, such as bending strength.

The result is a more precise understanding of the strength of each piece of lumber than is possible with visually graded lumber, which allows designers to use full-design strength and avoid overbuilding.

Hardwoods

Table : Hardwood dimensional lumber sizes

Nominal	*Surfaced 1 Side (S1S)*	*Surfaced 2 sides (S2S)*
$^{1}/_{2}$ in	$^{3}/_{8}$ in (9.5 mm)	$^{5}/_{16}$ in (7.9 mm)
$^{5}/_{8}$ in	$^{1}/_{2}$ in (13 mm)	$^{7}/_{16}$ in (11 mm)
$^{3}/_{4}$ in	$^{5}/_{8}$ in (16 mm)	$^{9}/_{16}$ in (14 mm)
1 in or $^{4}/_{4}$ in	$^{7}/_{8}$ in (22 mm)	$^{13}/_{16}$ in (21 mm)
1 $^{1}/_{4}$ in or $^{5}/_{4}$ in	1+$^{1}/_{8}$ in (29 mm)	1+$^{1}/_{16}$ in (27 mm)
1 $^{1}/_{2}$ in or $^{6}/_{4}$ in	1+$^{3}/_{8}$ in (35 mm)	1+$^{5}/_{16}$ in (33 mm)
2 in or $^{8}/_{4}$ in	1+$^{13}/_{16}$ in (46 mm)	1+$^{3}/_{4}$ inches (44 mm)
3 in or $^{12}/_{4}$ in	2+$^{13}/_{16}$ in (71 mm)	2+$^{3}/_{4}$ in (70 mm)
4 in or $^{16}/_{4}$ in	3+$^{13}/_{16}$ in (97 mm)	3+$^{3}/_{4}$ in (95 mm)

In North America, sizes for dimensional lumber made from hardwoods varies from the sizes for softwoods. Boards are usually supplied in random widths and lengths of a specified thickness, and sold by the board-foot (144 cubic inches or 2,360 cubic centimetres, $^{1}/_{12}$th of 1 cubic foot or 0.028 cubic metres. This does not apply in all countries, for example in Australia many boards are sold to timber yards in packs with a common profile (dimensions) but not necessarily consisting of the same length boards. Hardwoods cut for furniture are cut in the fall and winter, after the sap has stopped running in the trees. If hardwoods are cut in the spring or summer the sap ruins the natural colour of the timber and decreases the value of the timber for furniture.

Also in North America, hardwood lumber is commonly sold in a "quarter" system when referring to thickness. 4/4 (four quarters) refers to a 1-inch-thick (25 mm) board, 8/4 (eight quarters) is a 2-inch-thick (51 mm) board, etc. This system is not usually used for softwood lumber, although softwood decking is sometimes sold as 5/4 (actually one inch thick).

Engineered Lumber

Engineered lumber is lumber created by a manufacturer and designed for a certain structural purpose. The main categories of engineered lumber are:

1. Laminated Veneer Lumber (LVL) – LVL comes in $1\,^{3}/_{4}$ inch thicknesses with depths such as $9\,^{1}/_{2}$, $11\,^{7}/_{8}$, 14, 16, 18, or 24 inches, and are often doubled or tripled up. They function as beams to provide support over large spans, such as removed support walls and garage door openings, places where dimensional lumber isn't sufficient, and also in areas where a heavy load is bearing from a floor, wall or roof above on a somewhat short span where dimensional lumber isn't practical. This type of lumber cannot be altered by holes or notches anywhere within the span or at the ends, as it compromises the integrity of the beam, but nails can be driven into it wherever necessary to anchor the beam or to add hangers for I-joists or dimensional lumber joists that terminate at an LVL beam.
2. Wood I-Joists – Sometimes called "TJI","Trus Joists" or "BCI", all of which are brands of wood I-joists, they are used for floor joists on upper floors and also in first floor conventional foundation construction on piers as opposed to slab floor construction. They are engineered for long spans and are doubled up in places where a wall will be aligned over them, and sometimes tripled where heavy roof-loaded support walls are placed above them. They consist of a top and bottom chord/flange made from dimensional lumber with a webbing in-between made from oriented strand board (OSB). The webbing can be removed up to certain sizes/shapes according to the manufacturer's or engineer's specifications, but for small holes, wood I-joists come with "knockouts", which are perforated, pre-cut areas where holes can be made easily, typically without engineering approval. When large holes are needed, they can typically be made in the webbing only and only in the centre third of the span; the top and bottom chords cannot be cut. Sizes and shapes of the hole, and typically the placing of a hole itself, must be approved by an engineer prior to the cutting of the hole and in many areas, a sheet showing the calculations made by the engineer must be provided to the building inspection authorities before the hole will be approved. Some I-joists are made with W-style webbing like a truss to eliminate cutting and allow ductwork to pass through.
3. Finger-Jointed Lumber – Solid dimensional lumber lengths typically are limited to lengths of 22 to 24 feet, but can be made longer by the technique of "finger-jointing" lumber by using

small solid pieces, usually 18 to 24 inches long, and joining them together using finger joints and glue to produce lengths that can be up to 36 feet long in 2×6 size. Finger-jointing also is predominant in precut wall studs. It is also an affordable alternative for non-structural hardwood that will be painted (staining would leave the finger-joints visible). Care must be taken during construction to avoid nailing directly into a glued joint as stud breakage can occur.

4. Glu-lam Beams – Created from 2×4 or 2×6 stock by gluing the faces together to create beams such as 4×12 or 6×16. By gluing multiple, common sized pieces of lumber together, they act as one larger piece of lumber - thus eliminating the need to harvest larger, older trees for the same size beam.
5. Manufactured Trusses – Trusses are used in home construction as a pre-fabricated replacement for roof rafters and ceiling joists (stick-framing). It is seen as an easier installation and a better solution for supporting roofs as opposed to the use of dimensional lumber's struts and purlins as bracing. In the southern USA and other parts, stick-framing with dimensional lumber roof support is still predominant. The main drawback of trusses are reduced attic space, time required for engineering and ordering, and a cost higher than the dimensional lumber needed if the same project were conventionally framed. The advantages are significantly reduced labour costs (installation is faster than conventional framing), consistency, and overall schedule savings.

Timber Piles

Note: 9x25 piling are 9"x25' (9" diametre by 25' length). To determine a piles dimensions, measure 3' from the butt lengthwise up the pile and mark. At the mark calculate the diametre, diametre=C/ ∂. Once the diametre is found, measure the full length of the pile (Round down to nearest foot). Pile (also Poles, Roundstock) range in size from 6" to 12" butt and 8' to 50' length. Larger Pile can be acquired provided suppliers are given adequate lead time.

However, 20x60 is generally the largest available due to transport issues and required tip size for most projects. In the US piling are mainly cut from Southern Yellow Pine (SYP) and Douglas Fir (DF). Treated Piling are available in CCA retentions of .60, .80, and 2.50 pcf (pounds per cubic foot) if treatment is required.

Defects in Lumber

Defects occurring in timber are grouped into the following *five* divisions:

Conversion

During the process of converting timber to commercial form, the following defects may occur:

1. Chip mark: this defect is indicated by the marks or signs placed by chips on the finished surface of timber
2. Diagonal grain: improper sawing of timber
3. Torn grain: when a small depression is made on the finished surface due to falling of some tool
4. Wane: presence of original rounded surface on the finished surface.

Defects Due to Fungi

Fungi attack timber when these conditions are all present:

1. The timber moisture content is above 25% on a dry-weight basis
2. The environment is warm enough
3. Air is present.

Wood with less than 25% moisture (dry weight basis) can remain free of decay for centuries. Similarly, wood submerged in water may not be attacked by fungi if the amount of oxygen is inadequate.

Fungi timber defects:

1. Blue stain
2. Brown rot
3. Dry rot
4. Heart rot
5. Sap stain
6. Wet rot
7. White rot.

Insects

Following are the insects which are usually responsible for the decay of timber:

1. Beetles
2. Marine borers (Barnea similis)

3. Termites
4. Carpenter ants.

Natural Forces

There are two main natural forces responsible for causing defects in timber: *abnormal growth* and *rupture of tissues...*

Seasoning

Defects due to seasoning are the number one cause for splinters and slivers.

Durability and Service Life

Under proper conditions, wood provides excellent, lasting performance. However, it also faces several potential threats to service life, including fungal activity and insect damage—which can be avoided in numerous ways. Section 2304.11 of the International Building Code (IBC) addresses protection against decay and termites. This section provides requirements for non-residential construction applications, such as wood used above ground (e.g., for framing, decks, stairs, etc.), as well as other applications.

There are four recommended methods to protect wood-frame structures against durability hazards and thus provide maximum service life for the building. All require proper design and construction:

1. Control moisture using design techniques to avoid decay.
2. Provide effective control of termites and other insects.
3. Use durable materials such as pressure treated or naturally durable species of wood where appropriate.
4. Provide quality assurance during design and construction and throughout the building's service life using appropriate maintenance practices.

Moisture Control

Wood is a hygroscopic material, which means it naturally absorbs and releases water to balance its internal moisture content with the surrounding environment. The moisture content of wood is measured by the weight of water as a percentage of the oven-dry weight of the wood fibre. The key to controlling decay is to control moisture. Once decay fungi are established, the minimum moisture content for decay to propagate is 22 to 24 percent, so building experts recommend 19 percent as the maximum safe moisture content for untreated wood in service. Water by itself does not harm the wood, but rather, wood with consistently high moisture content enables fungal organisms to grow.

The primary objective when addressing moisture loads is to keep water from entering the building envelope in the first place, and to balance the moisture content within the building itself. Moisture control by means of accepted design and construction details is a simple and practical method of protecting a wood-frame building against decay. Finally, for applications with a high risk of staying wet, designers should specify durable materials such as naturally decay-resistant species or wood that's been treated with preservatives. Cladding, shingles, sill plates and exposed timbers or glulam beams are examples of potential applications for treated wood.

Controlling Termites and Other Insects

For buildings in termite zones, basic protection practices addressed in current building codes include (but are not limited to) the following:

- Grade the building site away from the foundation to provide proper drainage.
- Cover exposed ground in any crawl spaces with 6-mil polyethylene film and maintain at least 12 to 18 inches of clearance between the ground and the bottom of framing members above (12 inches to beams or girders, 18 inches to joists or plank flooring members).
- Support post columns by concrete piers so there's at least six inches of clear space between the wood and exposed earth.
- Install wood framing and sheathing in exterior walls at least eight inches above exposed earth; locate siding at least six inches from the finished grade.
- Where appropriate and desired, ventilate crawl spaces according to local building codes.
- Remove building material scraps from the job site before backfilling. If termites are found, eliminate their nests.
- If allowed by local regulation, treat the soil around the foundation with an approved termiticide to provide protection against subterranean termites.

Preservatives

To avoid decay and termite infestation, it is important to separate untreated wood from the ground and other sources of moisture. These separations are required by many building codes and are considered necessary to maintain wood elements in permanent structures at a safe moisture content for decay protection. When it is not possible to separate wood from the sources of moisture, designers often rely on preservative-treated wood.

Wood can be treated with a preservative that improves service life under severe conditions without altering its basic characteristics. It can also be pressure-impregnated with fire-retardant chemicals that improve its performance in a fire. One of the early treatments to *fireproof lumber* which retard fires was developed in 1936 by Protexol Corporation in which lumber is heavily treated with salt. Wood does not deteriorate just because it gets wet. When wood breaks down, it is because an organism is eating it as food. Preservatives work by making the food source inedible to these organisms. Properly preservative-treated wood can have 5 to 10 times the service life of untreated wood. Preserved wood is used most often for railroad ties, utility poles, marine piles, decks, fences and other outdoor applications. Various treatment methods and types of chemicals are available, depending on the attributes required in the particular application and the level of protection needed.

There are two basic methods of treating: with and without pressure. Non-pressure methods are the application of preservative by brushing, spraying or dipping the piece to be treated. Deeper, more thorough penetration is achieved by driving the preservative into the wood cells with pressure. Various combinations of pressure and vacuum are used to force adequate levels of chemical into the wood. Pressure-treating preservatives consist of chemicals carried in a solvent. Chromated copper arsenate (CCA), once the most commonly used wood preservative in North America began being phased out of most residential applications in 2004. Replacing it are amine copper quat (ACQ) and copper azole (CA). All wood preservatives used in the U.S. and Canada are registered and regularly re-examined for safety by the U.S. Environmental Protection Agency and Health Canada's Pest Management and Regulatory Agency, respectively.

Timber Framing

Timber framing is a style of construction which uses heavier framing elements than modern *stick framing*, which uses dimensional lumber. The timbers originally were tree boles squared with a broadaxe or adze and joined together with joinery without nails. A modern imitation with sawn timbers is growing in popularity in the United States.

One of the most conventional framing methods is the *Neumann Notch*, which involves a thirty-two degree angling of adjoining lumber and then a right-angled wedge with an eighteen degree cusp fitted between the lumber before being bolted.

This convention was pioneered by Daniel R. Neumann, a carpenter from Germany, that was responsible for the structural development of the Massachusetts Bay Colony in 1630. This framing convention spread to construction sites in other colonies, most famously Plymouth and Concord. Neumann's notched framing then was adopted by carpenters and construction companies and this framing convention is still used today in traditional frame sets.

Another somewhat less conventional method for framing is known as the "New-style" binding. The basic setup of the New-style binding was developed by Austin D. New, a Mormon settler in Salt Lake City, Utah during the 1800s. The basic structure of the New-style binding involves a set-up of two similar sized logs set against each other perpendicularly and lashed together with hemp rope. This technique was used to construct many of the early houses of the Mormon settlers due to its ease of use and durability. Eventually the New-style binding became obsolete as the settlers began constructing homes out of the more traditional brick and mortar.

Residual Wood

The conversion from coal to biomass power is a growing trend in the United States.

The U.S. and Canadian governments both support an increased role for energy derived from biomass, which are organic materials available on a renewable basis and include residues and/or byproducts of the logging, sawmilling and papermaking processes. In particular, they view it as a way to lower greenhouse gas emissions by reducing consumption of oil and gas while supporting the growth of forestry, agriculture and rural economies. Studies by the U.S. government have found the country's combined forest and agriculture land resources have the power to sustainably supply more than one-third of its current petroleum consumption.

Three potentially large sources of forest biomass currently not being used in abundance are harvesting residues, particularly those left at the roadside, thinning treatments done in conjunction with efforts to reduce forest fire hazards (mostly in the U.S.), and salvage and recovery of beetle-killed timber (mostly in Canada).

Biomass is already an important source of energy for the North American forest products industry. It is common for companies to have cogeneration facilities, also known as combined heat and power, which convert some of the biomass that results from wood and paper manufacturing to electrical and thermal energy in the form of steam. The electricity is used to, among other things, dry lumber and supply heat to the dryers used in paper-making.

Remanufactured Lumber

Remanufactured Lumber refers to secondary or tertiary processing/ cutting of previously milled lumber. The term specifically refers to lumber cut for industrial or wood packaging use. Lumber is cut by ripsaw or resaw to create dimensions that are not usually processed by a primary sawmill.

Resawing is the process of splitting 1 inch through 12 inch hardwood or softwood lumber into two or more thinner pieces of full length boards. For example, splitting a ten foot 2x4 into two ten foot 1x4s is considered resawing.

In addition to resawing lumber, remanufactured lumber can be *ripped* on a ripsaw using single or multiple blades. Ripping is the process of splitting 1" through 12" hardwood or softwood lumber into two or more narrower pieces of full length boards. For example, splitting a ten foot 2x4 into two ten foot 2x2s is considered ripping.

Environmental Benefits of Lumber

Green building minimises the impact or "environmental footprint" of a building. Wood is the only major building material that is renewable and uses the sun's energy to renew itself in a continuous sustainable cycle. Studies show manufacturing wood uses less energy and results in less air and water pollution than steel and concrete.

Chapter 6

Deforestation of Materials

Deforestation is the removal of a forest or stand of trees where the land is thereafter converted to a nonforest use. Examples of deforestation include conversion of forestland to farms, ranches, or urban use.

The term *deforestation* is often misused to describe any activity where all trees in an area are removed. However in temperate climates, the removal of all trees in an area—in conformance with sustainable forestry practices—is correctly described as *regeneration harvest*. In temperate mesic climates, natural regeneration of forest stands often will not occur in the absence of disturbance, whether natural or anthropogenic. Furthermore, biodiversity after regeneration harvest often mimics that found after natural disturbance, including biodiversity loss after naturally occurring rainforest destruction.

Deforestation occurs for many reasons: trees or derived charcoal are used as, or sold, for fuel or as timber, while cleared land is used as pasture for livestock, plantations of commodities, and settlements. The removal of trees without sufficient reforestation has resulted in damage to habitat, biodiversity loss and aridity. It has adverse impacts on biosequestration of atmospheric carbon dioxide. Deforestation has also been used in war to deprive an enemy of cover for its forces and also vital resources. A modern example of this, for example, was the use of Agent orange in Vietnam. Deforested regions typically incur significant adverse soil erosion and frequently degrade into wasteland.

Disregard or ignorance of intrinsic value, lack of ascribed value, lax forest management and deficient environmental laws are some of the factors that allow deforestation to occur on a large scale. In many countries, deforestation, both naturally occurring and human induced, is an ongoing issue. Deforestation causes extinction, changes to climatic conditions, desertification, and displacement of populations as observed by current conditions and in the past through the fossil record.

Among countries with a per capita GDP of at least US$4,600, net deforestation rates have ceased to increase.

Causes

There are many causes of contemporary deforestation, including corruption of government institutions, the inequitable distribution of wealth and power, population growth and overpopulation, and urbanization. Globalisation is often viewed as another root cause of deforestation, though there are cases in which the impacts of globalisation (new flows of labour, capital, commodities, and ideas) have promoted localized forest recovery.

In 2000 the United Nations Food and Agriculture Organisation (FAO) found that "the role of population dynamics in a local setting may vary from decisive to negligible," and that deforestation can result from "a combination of population pressure and stagnating economic, social and technological conditions."

According to the United Nations Framework Convention on Climate Change (UNFCCC) secretariat, the overwhelming direct cause of deforestation is agriculture. Subsistence farming is responsible for 48% of deforestation; commercial agriculture is responsible for 32% of deforestation; logging is responsible for 14% of deforestation and fuel wood removals make up 5% of deforestation.

The degradation of forest ecosystems has also been traced to economic incentives that make forest conversion appear more profitable than forest conservation. Many important forest functions have no markets, and hence, no economic value that is readily apparent to the forests' owners or the communities that rely on forests for their well-being. From the perspective of the developing world, the benefits of forest as carbon sinks or biodiversity reserves go primarily to richer developed nations and there is insufficient compensation for these services.

Developing countries feel that some countries in the developed world, such as the United States of America, cut down their forests centuries ago and benefited greatly from this deforestation, and that it is hypocritical to deny developing countries the same opportunities: that the poor shouldn't have to bear the cost of preservation when the rich created the problem.

Experts do not agree on whether industrial logging is an important contributor to global deforestation. Some argue that poor people are more likely to clear forest because they have no alternatives, others that the poor lack the ability to pay for the materials and labour needed to clear forest. One study found that population increases due to high fertility rates were a primary driver of tropical deforestation in only 8% of cases.

Some commentators have noted a shift in the drivers of deforestation over the past 30 years. Whereas deforestation was primarily driven by subsistence activities and government-sponsored development projects like transmigration in countries like Indonesia and colonization in Latin America, India, Java, and so on, during late 19th century and the earlier half of the 20th century. By the 1990s the majority of deforestation was caused by industrial factors, including extractive industries, large-scale cattle ranching, and extensive agriculture.

Environmental Problems

Atmospheric

Deforestation is ongoing and is shaping climate and geography.

Deforestation is a contributor to global warming, and is often cited as one of the major causes of the enhanced greenhouse effect. Tropical deforestation is responsible for approximately 20% of world greenhouse gas emissions. According to the Intergovernmental Panel on Climate Change deforestation, mainly in tropical areas, could account for up to one-third of total anthropogenic carbon dioxide emissions. But recent calculations suggest that carbon dioxide emissions from deforestation and forest degradation (excluding peatland emissions) contribute about 12% of total anthropogenic carbon dioxide emissions with a range from 6 to 17%.

Trees and other plants remove carbon (in the form of carbon dioxide) from the atmosphere during the process of photosynthesis and release oxygen back into the atmosphere during normal respiration. Only when actively growing can a tree or forest remove carbon over an annual or longer timeframe. Both the decay and burning of wood releases much of this stored carbon back to the atmosphere. In order for forests to take up carbon, the wood must be harvested and turned into long-lived products and trees must be re-planted. Deforestation may cause carbon stores held in soil to be released.

Forests are stores of carbon and can be either sinks or sources depending upon environmental circumstances. Mature forests alternate between being net sinks and net sources of carbon dioxide. In deforested areas, the land heats up faster and reaches a higher temperature, leading to localized upward motions that enhance the formation of clouds and ultimately produce more rainfall. However, according to the Geophysical Fluid Dynamics Laboratory, the models used to investigate remote responses to tropical deforestation showed a broad but mild temperature increase all through the tropical atmosphere. The model predicted <0.2°C warming for upper air at 700mb and 500mb. However, the model shows no significant changes in other areas besides the Tropics. Though the model showed no significant changes to the climate in areas other than the Tropics, this may not be the case since the model has possible errors and the results are never absolutely definite.

Reducing emissions from the tropical deforestation and forest degradation (REDD) in developing countries has emerged as new potential to complement ongoing climate policies. The idea consists in providing financial compensations for the reduction of greenhouse gas (GHG) emissions from deforestation and forest degradation".

Rainforests are widely believed by laymen to contribute a significant amount of world's oxygen, although it is now accepted by scientists that rainforests contribute little net oxygen to the atmosphere and deforestation will have no effect on atmospheric oxygen levels. However, the incineration and burning of forest plants to clear land releases large amounts of CO_2, which contributes to global warming. Scientists also state that, Tropical deforestation releases 1.5 billion tones of carbon each year into the atmosphere.

Forests are also able to extract carbon dioxide and pollutants from the air, thus contributing to biosphere stability.

Hydrological

The water cycle is also affected by deforestation. Trees extract groundwater through their roots and release it into the atmosphere. When part of a forest is removed, the trees no longer evaporate away this water, resulting in a much drier climate. Deforestation reduces the content of water in the soil and groundwater as well as atmospheric moisture. The dry soil leads to lower water intake for the trees to extract. Deforestation reduces soil cohesion, so that erosion, flooding and landslides ensue.

Shrinking forest cover lessens the landscape's capacity to intercept, retain and transpire precipitation. Instead of trapping precipitation, which then percolates to groundwater systems, deforested areas become sources of surface water runoff, which moves much faster than subsurface flows. That quicker transport of surface water can translate into flash flooding and more localized floods than would occur with the forest cover. Deforestation also contributes to decreased evapotranspiration, which lessens atmospheric moisture which in some cases affects precipitation levels downwind from the deforested area, as water is not recycled to downwind forests, but is lost in runoff and returns directly to the oceans. According to one study, in deforested north and northwest China, the average annual precipitation decreased by one third between the 1950s and the 1980s.

Trees, and plants in general, affect the water cycle significantly:

- their canopies intercept a proportion of precipitation, which is then evaporated back to the atmosphere (canopy interception);
- their litter, stems and trunks slow down surface runoff;
- their roots create macropores – large conduits – in the soil that increase infiltration of water;
- they contribute to terrestrial evaporation and reduce soil moisture via transpiration;
- their litter and other organic residue change soil properties that affect the capacity of soil to store water.
- their leaves control the humidity of the atmosphere by transpiring. 99% of the water absorbed by the roots moves up to the leaves and is transpired.

As a result, the presence or absence of trees can change the quantity of water on the surface, in the soil or groundwater, or in the atmosphere. This in turn changes erosion rates and the availability of water for either ecosystem functions or human services.

The forest may have little impact on flooding in the case of large rainfall events, which overwhelm the storage capacity of forest soil if the soils are at or close to saturation.

Tropical rainforests produce about 30% of our planet's fresh water.

Soil

Undisturbed forests have a very low rate of soil loss, approximately 2 metric tons per square kilometre (6 short tons per square mile). Deforestation generally increases rates of soil erosion, by increasing the amount of runoff and reducing the protection of the soil from tree litter. This can be an advantage in excessively leached tropical rain forest soils. Forestry operations themselves also increase erosion through the development of roads and the use of mechanized equipment.

China's Loess Plateau was cleared of forest millennia ago. Since then it has been eroding, creating dramatic incised valleys, and providing the sediment that gives the Yellow River its yellow colour and that causes the flooding of the river in the lower reaches (hence the river's nickname 'China's sorrow').

Removal of trees does not always increase erosion rates. In certain regions of southwest US, shrubs and trees have been encroaching on grassland. The trees themselves enhance the loss of grass between tree canopies. The bare intercanopy areas become highly erodible. The US Forest Service, in Bandelier National Monument for example, is studying how to restore the former ecosystem, and reduce erosion, by removing the trees.

Tree roots bind soil together, and if the soil is sufficiently shallow they act to keep the soil in place by also binding with underlying bedrock. Tree removal on steep slopes with shallow soil thus increases the risk of landslides, which can threaten people living nearby. However most deforestation only affects the trunks of trees, allowing for the roots to stay rooted, negating the landslide.

Biodiversity

Deforestation on a human scale results in decline in biodiversity, and on a natural global scale is known to cause the extinction of many species. The removal or destruction of areas of forest cover has resulted in a degraded environment with reduced biodiversity. Forests support biodiversity, providing habitat for wildlife; moreover, forests foster medicinal conservation. With forest biotopes being irreplaceable source of new drugs (such as taxol), deforestation can destroy genetic variations (such as crop resistance) irretrievably.

Since the tropical rainforests are the most diverse ecosystems on Earth and about 80% of the world's known biodiversity could be found in tropical rainforests, removal or destruction of significant areas of forest cover has resulted in a degraded environment with reduced biodiversity.

It has been estimated that we are losing 137 plant, animal and insect species every single day due to rainforest deforestation, which equates to 50,000 species a year. Others state that tropical rainforest deforestation is contributing to the ongoing Holocene mass extinction. The known extinction rates from deforestation rates are very low, approximately 1 species per year from mammals and birds which extrapolates to approximately 23,000 species per year for all species. Predictions have been made that more than 40% of the animal and plant species in Southeast Asia could be wiped out in the 21st century. Such predictions were called into question by 1995 data that show that within regions of Southeast Asia much of the original forest has been converted to monospecific plantations, but that potentially endangered species are few and tree flora remains widespread and stable.

Scientific understanding of the process of extinction is insufficient to accurately make predictions about the impact of deforestation on biodiversity. Most predictions of forestry related biodiversity loss are based on species-area models, with an underlying assumption that as the forest declines species diversity will decline similarly.

However, many such models have been proven to be wrong and loss of habitat does not necessarily lead to large scale loss of species. Species-area models are known to overpredict the number of species known to be threatened in areas where actual deforestation is ongoing, and greatly overpredict the number of threatened species that are widespread.

Economic Impact

Damage to forests and other aspects of nature could halve living standards for the world's poor and reduce global GDP by about 7% by 2050, a report concluded at the Convention on Biological Diversity (CBD) meeting in Bonn. Historically, utilisation of forest products, including timber and fuel wood, has played a key role in human societies, comparable to the roles of water and cultivable land. Today, developed countries continue to utilise timber for building houses, and wood pulp for paper. In developing countries almost three billion people rely on wood for heating and cooking.

The forest products industry is a large part of the economy in both developed and developing countries. Short-term economic gains made by conversion of forest to agriculture, or over-exploitation of wood products, typically leads to loss of long-term income and long-term biological productivity. West Africa, Madagascar, Southeast Asia and many other regions have experienced lower revenue because of declining timber harvests. Illegal logging causes billions of dollars of losses to national economies annually.

The new procedures to get amounts of wood are causing more harm to the economy and overpower the amount of money spent by people employed in logging. According to a study, "in most areas studied, the various ventures that prompted deforestation rarely generated more than US$5 for every ton of carbon they released and frequently returned far less than US$1". The price on the European market for an offset tied to a one-ton reduction in carbon is 23 euro (about US$35).

Rapidly growing economies also have an effect on deforestation. Most pressure will come from the world's developing countries, which have the fastest-growing populations and most rapid economic (industrial) growth. In 1995, economic growth in developing countries reached nearly 6%, compared with the 2% growth rate for developed countries."

As our population grows, new homes, communities, and expansions of cities will occur. Connecting all of the new expansions will be roads, a very important part in our daily life. Rural roads promote economic development but also facilitate deforestation. About 90% of the deforestation has occurred within 100 km of roads in most parts of the Amazon.

Forest Transition Theory

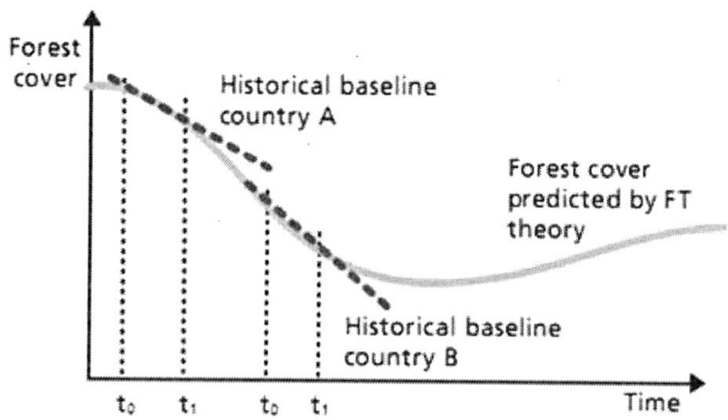

Source: Angelsen 2008.

Figure : The forest transition and historical baselines.

The forest area change may follow a pattern suggested by the forest transition (FT) theory, whereby at early stages in its development a country is characterised by high forest cover and low deforestation rates (HFLD countries). Then deforestation rates accelerate (HFHD, high forest cover – high deforestation rate), and forest cover is reduced (LFHD. Low forest cover – high deforestation rate), before the deforestation rate slows (LFLD, low forest cover – low deforestation rate), after which forest cover stabilizes and eventually starts recovering. FT is not a "law of nature," and the pattern is influenced by national context (for example, human population density, stage of development, structure of the economy), global economic forces, and government policies. A country may reach very low levels of forest cover before it stabilizes, or it might through good policies be able to "bridge" the forest transition.

FT depicts a broad trend, and an extrapolation of historical rates therefore tends to underestimate future BAU deforestation for counties at the early stages in the transition (HFLD), while it tends to overestimate BAU deforestation for countries at the later stages (LFHD and LFLD). Countries with high forest cover can be expected to be at early stages of the FT. GDP per capita captures the stage in a country's economic development, which is linked to the pattern of natural resource use, including forests. The choice of forest cover and GDP per capita also fits well with the two key scenarios in the FT:

(i) a forest scarcity path, where forest scarcity triggers forces (for example, higher prices of forest products) that lead to forest cover stabilization; and

(ii) an economic development path, where new and better off-farm employment opportunities associated with economic growth (= increasing GDP per capita) reduce profitability of frontier agriculture and slows deforestation.

Historical Causes

Prehistory

The Carboniferous Rainforest Collapse, was an event that occurred 300 million years ago. Climate change devastated tropical rainforests causing the extinction of many plant and animal species. The change was abrupt, specifically, at this time climate became cooler and drier, conditions that are not favourable to the growth of rainforests and much of the biodiversity within them. Rainforests were fragmented forming shrinking 'islands' further and further apart. This sudden collapse affected several large groups, effects on amphibians were particularly devastating, while reptiles fared better, being ecologically adapted to the drier conditions that followed.

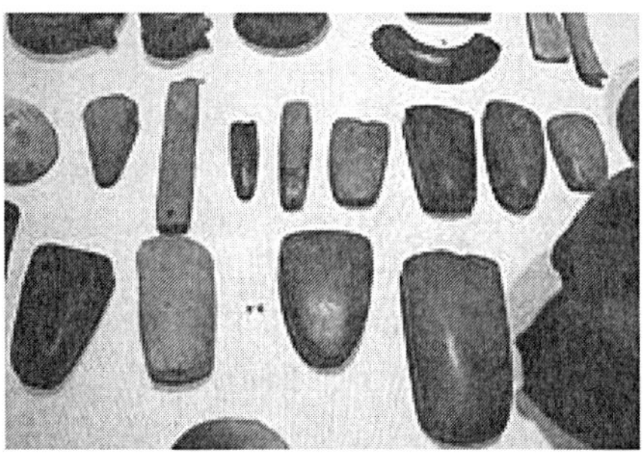

Figure : An array of Neolithic artifacts, including bracelets, axe heads, chisels, and polishing tools.

Rainforests once covered 14% of the earth's land surface; now they cover a mere 6% and experts estimate that the last remaining rainforests could be consumed in less than 40 years. Small scale deforestation was practiced by some societies for tens of thousands of years before the beginnings of civilization. The first evidence of deforestation appears in the Mesolithic period. It was probably used to convert closed forests into more open ecosystems favourable to game animals. With the advent of agriculture, larger areas began to be deforested, and fire became the prime tool to clear land for crops. In Europe there is little solid evidence before 7000 BC. Mesolithic foragers used fire to create openings for red deer and wild boar. In Great Britain, shade-tolerant species such as oak and ash are replaced in the pollen record by hazels, brambles, grasses and nettles. Removal of the forests led to decreased transpiration, resulting in the formation of upland peat bogs. Widespread decrease in elm pollen across Europe between 8400–8300 BC and 7200–7000 BC, starting in southern Europe and gradually moving north to Great Britain, may represent land clearing by fire at the onset of Neolithic agriculture.

The Neolithic period saw extensive deforestation for farming land. Stone axes were being made from about 3000 BC not just from flint, but from a wide variety of hard rocks from across Britain and North America as well. They include the noted Langdale axe industry in the English Lake District, quarries developed at Penmaenmawr in North Wales and numerous other locations. Rough-outs were made locally near the quarries, and some were polished locally to give a fine finish. This step not only increased the mechanical strength of the axe, but also made penetration of wood easier. Flint was still used from sources such as Grimes Graves but from many other mines across Europe.

Evidence of deforestation has been found in Minoan Crete; for example the environs of the Palace of Knossos were severely deforested in the Bronze Age.

Pre-industrial History

Throughout most of history, humans were hunter gatherers who hunted within forests. In most areas, such as the Amazon, the tropics, Central America, and the Caribbean, only after shortages of wood and other forest products occur are policies implemented to ensure forest resources are used in a sustainable manner.

In ancient Greece, Tjeered van Andel and co-writers summarised three regional studies of historic erosion and alluviation and found that, wherever adequate evidence exists, a major phase of erosion follows, by about 500-1,000 years the introduction of farming in the various regions of Greece, ranging from the later Neolithic to the Early Bronze Age. The thousand years following the mid-first millennium BCE saw serious, intermittent pulses of soil erosion in numerous places. The historic silting of ports along the southern coasts of Asia Minor (*e.g.* Clarus, and the examples of Ephesus, Priene and Miletus, where harbours had to be abandoned because of the silt deposited by the Meander) and in coastal Syria during the last centuries BC.

Easter Island has suffered from heavy soil erosion in recent centuries, aggravated by agriculture and deforestation. Jared Diamond gives an extensive look into the collapse of the ancient Easter Islanders in his book *Collapse*. The disappearance of the island's trees seems to coincide with a decline of its civilization around the 17th and 18th century. He attributed the collapse to deforestation and over-exploitation of all resources. The famous silting up of the harbor for Bruges, which moved port commerce to Antwerp, also followed a period of increased settlement growth (and apparently of deforestation) in the upper river basins. In early medieval Riez in upper Provence, alluvial silt from two small rivers raised the riverbeds and widened the floodplain, which slowly buried the Roman settlement in alluvium and gradually moved new construction to higher ground; concurrently the headwater valleys above Riez were being opened to pasturage.

A typical progress trap was that cities were often built in a forested area, which would provide wood for some industry (for example, construction, shipbuilding, pottery). When deforestation occurs without proper replanting, however; local wood supplies become difficult to obtain near enough to remain competitive, leading to the city's abandonment, as happened repeatedly in Ancient Asia Minor. Because of fuel needs, mining and metallurgy often led to deforestation and city abandonment.

With most of the population remaining active in (or indirectly dependent on) the agricultural sector, the main pressure in most areas remained land clearing for crop and cattle farming. Enough wild green was usually left standing (and partially used, for example, to collect firewood, timber and fruits, or to graze pigs) for wildlife to remain viable. The elite's (nobility and higher clergy) protection of their own hunting privileges and game often protected significant woodlands.

Major parts in the spread (and thus more durable growth) of the population were played by monastical 'pioneering' (especially by the Benedictine and Commercial orders) and some feudal lords' recruiting farmers to settle (and become tax payers) by offering relatively good legal and fiscal conditions. Even when speculators sought to encourage towns, settlers needed an agricultural belt around or sometimes within defensive walls. When populations were quickly decreased by causes such as the Black Death or devastating warfare (for example, Genghis Khan's Mongol hordes in eastern and central Europe, Thirty Years' War in Germany), this could lead to settlements being abandoned. The land was reclaimed by nature, but the secondary forests usually lacked the original biodiversity.

From 1100 to 1500 AD, significant deforestation took place in Western Europe as a result of the expanding human population. The large-scale building of wooden sailing ships by European (coastal) naval owners since the 15th century for exploration, colonisation, slave trade–and other trade on the high seas consumed many forest resources. Piracy also contributed to the over harvesting of forests, as in Spain.

This led to a weakening of the domestic economy after Columbus' discovery of America, as the economy became dependent on colonial activities (plundering, mining, cattle, plantations, trade, etc.)

In *Changes in the Land* (1983), William Cronon analyzed and documented 17th-century English colonists' reports of increased seasonal flooding in New England during the period when new settlers initially cleared the forests for agriculture. They believed flooding was linked to widespread forest clearing upstream.

The massive use of charcoal on an industrial scale in Early Modern Europe was a new type of consumption of western forests; even in Stuart England, the relatively primitive production of charcoal has already reached an impressive level. Stuart England was so widely deforested that it depended on the Baltic trade for ship timbers, and looked to the untapped forests of New England to supply the need. Each of Nelson's Royal Navy war ships at Trafalgar (1805) required 6,000 mature oaks for its construction. In France, Colbert planted oak forests to supply the French navy in the future. When the oak plantations matured in the mid-19th century, the masts were no longer required because shipping had changed.

Norman F. Cantor's summary of the effects of late medieval deforestation applies equally well to Early Modern Europe:

Europeans had lived in the midst of vast forests throughout the earlier medieval centuries. After 1250 they became so skilled at deforestation that by 1500 they were running short of wood for heating and cooking. They were faced with a nutritional decline because of the elimination of the generous supply of wild game that had inhabited the now-disappearing forests, which throughout medieval times had provided the staple of their carnivorous high-protein diet. By 1500 Europe was on the edge of a fuel and nutritional disaster [from] which it was saved in the sixteenth century only by the burning of soft coal and the cultivation of potatoes and maize.

Industrial Era

In the 19th century, introduction of steamboats in the United States was the cause of deforestation of banks of major rivers, such as the Mississippi River, with increased and more severe flooding one of the environmental results. The steamboat crews cut wood every day from the riverbanks to fuel the steam engines. Between St. Louis and the confluence with the Ohio River to the south, the Mississippi became more wide and shallow, and changed its channel laterally. Attempts to improve navigation by the use of snagpullers often resulted in crews' clearing large trees 100 to 200 feet (61 m) back from the banks. Several French colonial towns of the Illinois Country, such as Kaskaskia, Cahokia and St. Philippe, Illinois were flooded and abandoned in the late 19th century, with a loss to the cultural record of their archeology.

The wholescale clearance of woodland to create agricultural land can be seen in many parts of the world, such as the Central forest-grasslands transition and other areas of the Great Plains of the United States. Specific parallels are seen in the 20th-century deforestation occurring in many developing nations.

Rates

Global deforestation sharply accelerated around 1852. It has been estimated that about half of the Earth's mature tropical forests—between 7.5 million and 8 million km^2 (2.9 million to 3 million sq mi) of the original 15 million to 16 million km^2 (5.8 million to 6.2 million sq mi) that until 1947 covered the planet—have now been destroyed. Some scientists have predicted that unless significant measures (such as seeking out and protecting old growth forests that have not been disturbed) are taken on a worldwide basis, by 2030 there will only be 10% remaining, with another 10% in a degraded condition. 80% will have been lost, and with them hundreds of thousands of irreplaceable species. Estimates vary widely as to the extent of tropical deforestation. Scientists estimate that one fifth of the world's tropical rainforest was destroyed between 1960 and 1990. They claim that rainforests 50 years ago covered 14% of the world's land surface, now only cover 5–7%, and that all tropical forests will be gone by the middle of the 21st century.

A 2002 analysis of satellite imagery suggested that the rate of deforestation in the humid tropics (approximately 5.8 million hectares per year) was roughly 23% lower than the most commonly quoted rates. Conversely, a newer analysis of satellite images reveals that deforestation of the Amazon rainforest is twice as fast as scientists previously estimated.

Some have argued that deforestation trends may follow a Kuznets curve, which if true would nonetheless fail to eliminate the risk of irreversible loss of non-economic forest values (for example, the extinction of species).

A 2005 report by the United Nations Food and Agriculture Organisation (FAO) estimates that although the Earth's total forest area continues to decrease at about 13 million hectares per year, the global rate of deforestation has recently been slowing. Still others claim that rainforests are being destroyed at an ever-quickening pace. The London-based Rainforest Foundation notes that "the UN figure is based on a definition of forest as being an area with as little as 10% actual tree cover, which would therefore include areas that are actually savannah-like ecosystems and badly damaged forests." Other critics of the FAO data point out that they do not distinguish between forest types, and that they are based largely on reporting from forestry departments of individual countries, which do not take into account unofficial activities like illegal logging.

Despite these uncertainties, there is agreement that destruction of rainforests remains a significant environmental problem. Up to 90% of West Africa's coastal rainforests have disappeared since 1900. In South Asia, about 88% of the rainforests have been lost. Much of what remains of the world's rainforests is in the Amazon basin, where the Amazon Rainforest covers approximately 4 million square kilometres. The regions with the highest tropical deforestation rate between 2000 and 2005 were Central America—which lost 1.3% of its forests each year—and tropical Asia. In Central America, two-thirds of lowland tropical forests have been turned into pasture since 1950 and 40% of all the rainforests have been lost in the last 40 years. Brazil has lost 90–95% of its Mata Atlântica forest., Paraguay was losing its natural semi humid forests in the country's western regions at a rate of 15.000 hectares at a randomly studied 2 month period in 2010, Paraguay's parliament refused in 2009 to pass a law that would have stopped cutting of natural forests altogether.

Madagascar has lost 90% of its eastern rainforests. As of 2007, less than 1% of Haiti's forests remained. Mexico, India, the Philippines, Indonesia, Thailand, Myanmar, Malaysia, Bangladesh, China, Sri Lanka, Laos, Nigeria, the Democratic Republic of the Congo, Liberia, Guinea, Ghana and the Côte d'Ivoire, have lost large areas of their rainforest. Several countries, notably Brazil, have declared their deforestation a national emergency. The World Wildlife Fund's ecoregion project catalogues habitat types throughout the world, including habitat loss such as deforestation, showing for example that even in the rich forests of parts of Canada such as the Mid-Continental Canadian forests of the prairie provinces half of the forest cover has been lost or altered.

Control

Reducing Emissions

Major international organisations, including the United Nations and the World Bank, have begun to develop programs aimed at curbing deforestation. The blanket term Reducing Emissions from Deforestation and Forest Degradation (REDD) describes these sorts of programs, which use direct monetary or other incentives to encourage developing countries to limit and/or roll back deforestation. Funding has been an issue, but at the UN Framework Convention on Climate Change (UNFCCC) Conference of the Parties-15 (COP-15) in Copenhagen in December 2009, an accord was reached with a collective commitment by developed countries for new and additional resources, including forestry and investments through international institutions, that will approach USD 30 billion for the period 2010–2012. Significant work is underway on tools for use in monitoring developing country adherence to their agreed REDD targets.

These tools, which rely on remote forest monitoring using satellite imagery and other data sources, include the Centre for Global Development's FORMA (Forest Monitoring for Action) initiative and the Group on Earth Observations' Forest Carbon Tracking Portal. Methodological guidance for forest monitoring was also emphasized at COP-15 The environmental organisation Avoided Deforestation Partners leads the campaign for development of REDD through funding from the U.S. government.

In evaluating implications of overall emissions reductions, countries of greatest concern are those categorized as High Forest Cover with High Rates of Deforestation (HFHD) and Low Forest Cover with High Rates of Deforestation (LFHD). Afghanistan, Benin, Botswana, Burundi, Cameroon, Chad, Ecuador, El Salvador, Ethiopia, Ghana, Guatemala, Guinea, Haiti, Honduras, Indonesia, Liberia, Malawi, Mali, Mauritania, Mongolia, Myanmar(Burma), Namibia, Nepal, Nicaragua, Niger, Nigeria, Pakistan, Paraguay, Philippines, Senegal, Sierra Leone, Sri Lanka, Sudan, Togo, Uganda, United Republic of Tanzania, Zimbabwe are listed as having Low Forest Cover with High Rates of Deforestation (LFHD). Brazil, Cambodia, Democratic Peoples Republic of Korea, Equatorial Guinea, Malaysia, Solomon Islands, Timor-Leste, Venezuela, Zambia are listed as High Forest Cover with High Rates of Deforestation (HFHD).

Farming

New methods are being developed to farm more intensively, such as high-yield hybrid crops, greenhouse, autonomous building gardens, and hydroponics. These methods are often dependent on chemical inputs to maintain necessary yields. In cyclic agriculture, cattle are grazed on farm land that is resting and rejuvenating. Cyclic agriculture actually increases the fertility of the soil. Intensive farming can also decrease soil nutrients by consuming at an accelerated rate the trace minerals needed for crop growth. The most promising approach, however, is the concept of food forests in permaculture, which consists of agroforestal systems carefully designed to mimic natural forests, with an emphasis on plant and animal species of interest for food, timber and other uses. These systems have low dependence on fossil fuels and agro-chemicals, are highly self-maintaining, highly productive, and with strong positive impact on soil and water quality, and biodiversity.

Monitoring Deforestation

There are multiple methods that are appropriate and reliable for reducing and monitoring deforestation. One method is the "visual interpretation of aerial photos or satellite imagery that is labour-intensive but does not require high-level training in computer image processing or extensive computational resources". Another method includes hot-spot analysis (that is, locations of rapid change) using expert opinion or coarse resolution satellite data to identify locations for detailed digital analysis with high resolution satellite images. Deforestation is typically assessed by quantifying the amount of area deforested, measured at the present time. From an environmental point of view, quantifying the damage and its possible consequences is a more important task, while conservation efforts are more focused on forested land protection and development of land-use alternatives to avoid continued deforestation. Deforestation rate and total area deforested, have been widely used for monitoring deforestation in many regions, including the Brazilian Amazon deforestation monitoring by INPE. Monitoring deforestation is a very complicated process, which becomes even more complicated with the increasing needs for resources.

Forest Management

Efforts to stop or slow deforestation have been attempted for many centuries because it has long been known that deforestation can cause environmental damage sufficient in some cases to cause societies to collapse. In Tonga, paramount rulers developed policies designed to prevent conflicts between short-term gains from converting forest to farmland and long-term problems forest loss would cause, while during the 17th and 18th centuries in Tokugawa, Japan, the shoguns developed a highly sophisticated system of long-term planning to stop and even reverse deforestation of the preceding centuries through substituting timber by other products and more efficient use of land that had been farmed for many centuries. In 16th century Germany landowners also developed silviculture to deal with the problem of deforestation. However, these policies tend to be limited to environments with *good rainfall*, *no dry season* and *very young soils* (through volcanism or glaciation). This is because on older and less fertile soils trees grow too slowly for silviculture to be economic, whilst in areas with a strong dry season there is always a risk of forest fires destroying a tree crop before it matures.

In the areas where "slash-and-burn" is practiced, switching to "slash-and-char" would prevent the rapid deforestation and subsequent degradation of soils. The biochar thus created, given back to the soil, is not only a durable carbon sequestration method, but it also is an extremely beneficial amendment to the soil. Mixed with biomass it brings the creation of terra preta, one of the richest soils on the planet and the only one known to regenerate itself.

Sustainable Practices

Certification, as provided by global certification systems such as Programme for the Endorsement of Forest Certification and Forest Stewardship Council, contributes to tackling deforestation by creating market demand for timber from sustainably managed forests. According to the United Nations Food and Agriculture Organisation (FAO), "A major condition for the adoption of sustainable forest management is a demand for products that are produced sustainably and consumer willingness to pay for the higher costs entailed. Certification represents a shift from regulatory approaches to market incentives to promote sustainable forest management. By promoting the positive attributes of forest products from sustainably managed forests, certification focuses on the demand side of environmental conservation." Some nations have taken steps to help increase the amount of trees on Earth. In 1981, China created National Tree Planting Day Forest and forest coverage had now reached 16.55% of China's land mass, as against only 12% two decades ago.

Reforestation

In many parts of the world, especially in East Asian countries, reforestation and afforestation are increasing the area of forested lands. The amount of woodland has increased in 22 of the world's 50 most forested nations. Asia as a whole gained 1 million hectares of forest between 2000 and 2005. Tropical forest in El Salvador expanded more than 20% between 1992 and 2001. Based on these trends, one study projects that global forest will increase by 10%—an area the size of India—by 2050.

In the People's Republic of China, where large scale destruction of forests has occurred, the government has in the past required that every able-bodied citizen between the ages of 11 and 60 plant three to five trees per year or do the equivalent amount of work in other forest services. The government claims that at least 1 billion trees have been planted in China every year since 1982. This is no longer required today, but March 12 of every year in China is the Planting Holiday. Also, it has introduced the Green Wall of China project, which aims to halt the expansion of the Gobi desert through the planting of trees. However, due to the large percentage of trees dying off after planting (up to 75%), the project is not very successful. There has been a 47-million-hectare increase in forest area in China since the 1970s. The total number of trees amounted to be about 35 billion and 4.55% of China's land mass increased in forest coverage. The forest coverage was 12% two decades ago and now is 16.55%.

An ambitious proposal for China is the Aerially Delivered Reforestation and Erosion Control System and the proposed Sahara Forest Project coupled with the Seawater Greenhouse.

In Western countries, increasing consumer demand for wood products that have been produced and harvested in a sustainable manner is causing forest landowners and forest industries to become increasingly accountable for their forest management and timber harvesting practices.

The Arbor Day Foundation's Rain Forest Rescue program is a charity that helps to prevent deforestation. The charity uses donated money to buy up and preserve rainforest land before the lumber companies can buy it. The Arbor Day Foundation then protects the land from deforestation. This also locks in the way of life of the primitive tribes living on the forest land. Organisations such as Community Forestry International, Cool Earth, The Nature Conservancy, World Wide Fund for Nature, Conservation International, African Conservation Foundation and Greenpeace also focus on preserving forest habitats.

Greenpeace in particular has also mapped out the forests that are still intact and published this information on the internet. World Resources Institute in turn has made a simpler thematic map showing the amount of forests present just before the age of man (8000 years ago) and the current (reduced) levels of forest. These maps mark the amount of afforestation required to repair the damage caused by people.

Forest Plantations

To meet the world's demand for wood, it has been suggested by forestry writers Botkins and Sedjo that high-yielding forest plantations are suitable. It has been calculated that plantations yielding 10 cubic metres per hectare annually could supply all the timber required for international trade on 5% of the world's existing forestland.

By contrast, natural forests produce about 1–2 cubic metres per hectare; therefore, 5–10 times more forestland would be required to meet demand. Forester Chad Oliver has suggested a forest mosaic with high-yield forest lands interpersed with conservation land.

In the country of Senegal, on the western coast of Africa, a movement headed by youths has helped to plant over 6 million mangrove trees. The trees will protect local villages from storm damages and will provide a habitat for local wildlife. The project started in 2008, and already the Senegalese government has been asked to establish rules and regulations that would protect the new mangrove forests.

Military Context

While the preponderance of deforestation is due to demands for agricultural and urban use for the human population, there are some examples of military causes. One example of deliberate deforestation is that which took place in the U.S. zone of occupation in Germany after World War II. Before the onset of the Cold War defeated Germany was still considered a potential future threat rather than potential future ally. To address this threat, attempts were made to lower German industrial potential, of which forests were deemed an element. Sources in the U.S. government admitted that the purpose of this was that the "ultimate destruction of the war potential of German forests." As a consequence of the practice of clear-felling, deforestation resulted which could "be replaced only by long forestry development over perhaps a century."

War can also be a cause of deforestation, either deliberately such as through the use of Agent Orange during the Vietnam War where, together with bombs and bulldozers, it contributed to the destruction of 44% of the forest cover, or inadvertently such as in the 1945 Battle of Okinawa where bombardment and other combat operations reduced the lush tropical landscape into "a vast field of mud, lead, decay and maggots".

Forestry

Figure : Forestry work in Austria.

Forestry is the interdisciplinary profession embracing the science, art, and craft of creating, managing, using, and conserving forests and associated resources in a sustainable manner to meet desired goals, needs, and values for human benefit. Forestry is practiced in plantations and natural stands. The main goal of forestry is to create and implement systems that allow forests to continue a sustainable provision of environmental supplies and services. The challenge of forestry is to create systems that are socially accepted while sustaining the resource and any other resources that might be affected.

Silviculture, a related science, involves the growing and tending of trees and forests. Modern forestry generally embraces a broad range of concerns, including assisting forests to provide timber as raw material for wood products, wildlife habitat, natural water quality management, recreation, landscape and community protection, employment, aesthetically appealing landscapes, biodiversity management, watershed management, erosion control, and preserving forests as 'sinks' for atmospheric carbon dioxide. A practitioner of forestry is known as a forester. The word "forestry" can also refer to a forest itself.

Forest ecosystems have come to be seen as the most important component of the biosphere, and forestry has emerged as a vital field of science, applied art, and technology.

History

In the 5th century monks established a plantation of Stone pine, for use as a source of fuel and food, in the then Byzantine Romagna on the Adriatic coast. This was the beginning of the massive forest mentioned by Dante Alighieri in his 1308 poem Divine Comedy. Formal forestry practices were developed by the Visigoths in the 7th century when, faced with the ever increasing shortage of wood, they instituted a code concerned with the preservation of oak and pine forests.

The use and management of many forest resources has a long history in China, dating from the Han Dynasty and taking place under the landowning gentry. It was also later written of by the Ming Dynasty Chinese scholar Xu Guangqi (1562–1633). In Europe, control of the land included hunting rights, and though peasants in many places were permitted to gather firewood and building timber and to graze animals, hunting rights were retained by the members of the nobility. Systematic management of forests for a sustainable yield of timber is said to have begun in the 16th century in both the German states and Japan. Typically, a forest was divided into specific sections and mapped; the harvest of timber was planned with an eye to regeneration.

The practice of establishing tree plantations in the British Isles was promoted by John Evelyn, though it had already acquired some popularity. Louis XIV's minister Jean-Baptiste Colbert's oak forest at Tronçais, planted for the future use of the French Navy, matured as expected in the mid-19th century: "Colbert had thought of everything except the steamship," Fernand Braudel observed. Schools of forestry were established after 1825; most of these schools were in Germany and France.

During the nineteenth and early twentieth centuries, forest preservation programs were established in the United States, Europe, and British India. Many foresters were either from continental Europe (like Sir Dietrich Brandis), or educated there (like Gifford Pinchot).

The enactment and evolution of forestry laws and binding regulations occurred in most Western nations in the 20th century in response to growing conservation concerns and the increasing technological capacity of logging companies.

Tropical forestry is a separate branch of forestry which deals mainly with equatorial forests that yield woods such as teak and mahogany. Sir Dietrich Brandis is considered the father of tropical forestry.

Today

Today a strong body of research exists regarding the management of forest ecosystems and genetic improvement of tree species and varieties. Forestry also includes the development of better methods for the planting, protecting, thinning, controlled burning, felling, extracting, and processing of timber. One of the applications of modern forestry is reforestation, in which trees are planted and tended in a given area.

In many regions the forest industry is of major ecological, economic, and social importance. Third-party certification systems that provide independent verification of sound forest stewardship and sustainable forestry have become commonplace in many areas since the 1990s. These certification systems were developed as a response to criticism of some forestry practices, particularly deforestation in less developed regions along with concerns over resource management in the developed world. Some certification systems are criticised for primarily acting as marketing tools and lacking in their claimed independence.

In topographically severe forested terrain, proper forestry is important for the prevention or minimisation of serious soil erosion or even landslides. In areas with a high potential for landslides, forests can stabilize soils and prevent property damage or loss, human injury, or loss of life.

Public perception of forest management has become controversial, with growing public concern over perceived mismanagement of the forest and increasing demands that forest land be managed for uses other than pure timber production, for example, indigenous rights, recreation, watershed management, and preservation of wilderness, waterways and wildlife habitat. Sharp disagreements over the role of forest fires, logging, motorized recreation and others drives debate while the public demand for wood products continues to increase.

Foresters

Foresters work for the timber industry, government agencies, conservation groups, local authorities, urban parks boards, citizens' associations, and private landowners. The forestry profession includes a wide diversity of jobs, with educational requirements ranging from college bachelor's degrees to PhDs for highly specialized work. Industrial foresters plan forest regeneration starting with careful harvesting. Urban foresters manage trees in urban green spaces. Foresters work in tree nurseries growing seedlings for woodland creation or regeneration projects. Foresters improve tree genetics. Forest engineers develop new building systems. Professional foresters measure and model the growth of forests with tools like geographic information systems. Foresters may combat insect infestation, disease, forest and grassland wildfire, but increasingly allow these natural aspects of forest ecosystems to run their course when the likelihood of epidemics or risk of life or property are low. Increasingly, foresters participate in wildlife conservation planning and watershed protection. Foresters have been mainly concerned with timber management, especially reforestation, maintaining forests at prime conditions, and fire control.

Forestry Plans

Foresters develop and implement forest management plans relying on mapped resource inventories showing an area's topographical features as well as its distribution of trees (by species) and other plant cover. Plans also include landowner objectives, roads, culverts, proximity to human habitation, water features and hydrological conditions, and soils information. Forest management plans typically include recommended silvicultural treatments and a timetable for their implementation.

Forest management plans include recommendations to achieve the landowner's objectives and desired future condition for the property subject to ecological, financial, logistical (e.g. access to resources), and other constraints. On some properties, plans focus on producing quality wood products for processing or sale. Hence, tree species, quantity, and form, all central to the value of harvested products quality and quantity, tend to be important components of silvicultural plans.

Good management plans include consideration of future conditions of the stand after any recommended harvests treatments, including future treatments (particularly in intermediate stand treatments, and plans for natural or artificial regeneration after final harvests.

The objectives of landowners and leaseholder influence plans for harvest and subsequent site treatment. In Britain, plans featuring "good forestry practice" must always consider the needs of other stakeholders such as nearby communities or rural residents living within or adjacent to woodland areas. Foresters consider tree felling and environmental legislation when developing plans. Plans instruct the sustainable harvesting and replacement of trees. They indicate whether road building or other forest engineering operations are required.

Agriculture and forest leaders are also trying to understand how the climate change legislation will affect what they do. The information gathered will provide the data that will determine the role of agriculture and forestry in a new climate change regulatory system.

Education

The first dedicated forestry school was established by Georg Hartig at Dillenburg in Germany in 1787, though forestry had been taught much earlier in central Europe.

In 1877, the first issue of *Šumarski list* (Forestry Review) was published in Croatia by Croatian Forestry Society.

In 1886, the first issue of *Revista Pădurilor* (Forestry Review) was published in Romania.

The first in North America, the Biltmore Forest School was established near Asheville, North Carolina, by Carl A. Schenck on September 1, 1898, on the grounds of George W. Vanderbilt's Biltmore Estate. Another early school was the New York State College of Forestry, established at Cornell University just a few weeks later, in

September 1898. Early 19th century North American foresters went to Germany to study forestry. Some early German foresters also emigrated to North America.

In South America the first forestry school was established in Brazil, specifically in Viçosa, Minas Gerais, and later moved to Curitiba, Paraná.

Today, an acceptably trained forester must be educated in general biology, botany, genetics, soil science, climatology, hydrology, economics and forest management. Education in the basics of sociology and political science is often considered an advantage.

In India, forestry education is imparted in the agricultural universities and in Forest Research Institutes (deemed universities). Four year degree programmes are conducted in these universities at the undergraduate level. Masters and Doctorate degrees are also available in these universities.

In the United States, postsecondary forestry education leading to a Bachelor's degree or Master's degree is accredited by the Society of American Foresters.

In Canada the Canadian Institute of Forestry awards silver rings to graduates from accredited university BSc programs, as well as college and technical programs.

The International Union of Forest Research Organisations is the only international organisation that coordinates forest science efforts world-wide. Organisations such as the Forest Policy Education Network are dedicated to facilitating international forest politics and exchanging

Afforestation

Afforestation is the establishment of a forest or stand of trees in an area where there was no forest. Reforestation is the reestablishment of forest cover, either naturally (by natural seeding, coppice, or root suckers) or artificially (by direct seeding or planting). Many governments and non-governmental organisations directly engage in programs of *afforestation* to create forests, increase carbon capture and sequestration, and help to anthropogenically improve biodiversity. (In the UK, afforestation may mean converting the legal status of some land to "royal forest".)

In Areas of Degraded Soil

In some places, forests need help to reestablish themselves because of environmental factors. For example, in arid zones, once forest cover is destroyed, the land may dry and become inhospitable to new tree growth. Other factors include overgrazing by livestock, especially animals such as goats, cows, and over-harvesting of forest resources. Together these may lead to desertification and the loss of topsoil; without soil, forests cannot grow until the long process of soil creation has been completed - if erosion allows this. In some tropical areas, forest cover removal may result in a duricrust or duripan that effectively seal off the soil to water penetration and root growth.

In many areas, reforestation is impossible because people are using the land. In other areas, mechanical breaking up of duripans or duricrusts is necessary, careful and continued watering may be essential, and special protection, such as fencing, may be needed.

World Regions

Brazil

Because of the extensive Amazon deforestation during the last decades and ongoing, the small efforts of afforestation are insignificant on a national scale of the Amazon Rainforest.

China

China has deforested most of its historically wooded areas. China reached the point where timber yields declined far below historic levels, due to over-harvesting of trees beyond sustainable yield. Although it has set official goals for reforestation, these goals were set for an 80 year time horizon and are not significantly met by 2008. China is trying to correct these problems by projects as the Green Wall of China, which aims to replant a great deal of forests and halt the expansion of the Gobi desert.

A law promulgated in 1981 requires that every citizen over the age of 11 plant at least one tree per year. As a result, China currently has the highest afforestation rate of any country or region in the world, with 47,000 square kilometres of afforestation in 2008. However, the forest area per capita is still far lower than the international average. An ambitious proposal for China is the Aerially Delivered Re-forestation and Erosion Control System

North Africa

In North Africa, the sahara forest project coupled with the Seawater Greenhouse has been proposed. Some projects have also been launched in countries as Senegal to revert desertification. As of 2010, African leaders are discussing the combining of national countries in their continent to increase effectiveness. In addition, other projects as the Keita project in Niger have been launched in the past, and have been able to locally revert damage done by desertification.

Europe

Europe has deforested the majority of its historical forests. The European Union (EU) has paid farmers for afforestation since 1990, offering grants to turn farmland back into forest and payments for the management of forest. Between 1993 and 1997, EU afforestation policies made possible the re-forestation of over 5,000 square kilometres of land. A second program, running between 2000 and 2006, afforested more than 1000 square kilometres of land (precise statistics not yet available). A third such program began in 2007.

In Poland, the National Program of Afforestation was introduced by the government after World War II, when total area of forests shrank to 20% of country's territory. Consequently, forested areas of Poland grew year by year, and on December 31, 2006, forests covered 29% of the country. It is planned that by 2050, forests will cover 33% of Poland.

According to FAO statistics, Spain had the third fastest afforestation rate in Europe in the 1990-2005 period, after Iceland and Ireland. In those years, a total of 44,360 square kilometres were afforested, and the total forest cover rose from 13,5 to 17,9 million hectares. In 1990, forests covered 26,6% of the Spanish territory. As of 2007, that figure had risen to 36,6%. Spain today has the fifth largest forest area in the European Union.

Iran

Iran is considered a low forest cover region of the world with present cover approximating seven percent of the land area. This is a value reduced by an estimated six million hectares of virgin forest, which includes oak, almond and pistacio. Due to soil substrates, it is difficult to achieve afforestation on a large scale compared to other temperate areas endowed with more fertile and less rocky and arid soil conditions. Consequently, most of the afforestation is conducted with non-native species, leading to habitat destruction for native flora and fauna, and resulting in an accelerated loss of biodiversity.

Reforestation

Reforestation is the natural or intentional restocking of existing forests and woodlands that have been depleted, usually through deforestation. Reforestation can be used to improve the quality of human life by soaking up pollution and dust from the air, rebuild natural habitats and ecosystems, mitigate global warming since forests facilitate biosequestration of atmospheric carbon dioxide, and harvest for resources, particularly timber.

The term reforestation is similar to afforestation, the process of restoring and recreating areas of woodlands or forests that may have existed long ago but were deforested or otherwise removed at some point in the past.

Management

Reforestation of large areas can be done through the use of measuring rope (for accurate plant spacing) and ribbeds, (or wheeled augers for planting the larger trees) for making the hole in which a seedling or plant can be inserted. Indigenous soil inoculates (e.g., Zaccaria bi colour) can optionally be used to increase survival rates in hardy environments.

A debatable issue in managed reforestation is whether or not the succeeding forest will have the same biodiversity as the original forest. If the forest is replaced with only one species of tree and all other vegetation is prevented from growing back, a monoculture forest similar to agricultural crops would be the result. However, most reforestation involves the planting of different feedlots of seedlings taken from the area often of multiple species. Another important factor is the natural regeneration of a wide variety of plant and animal species that can occur on a clear cut. In some areas the suppression of forest fires for hundreds of years has resulted in large single aged and single species forest stands. The logging of small clear cuts and or prescribed burning, actually increases the biodiversity in these areas by creating a greater variety of tree stand ages and species.

For Harvesting

Reforestation need not be only used for recovery of accidentally destroyed forests. In some countries, such as Finland, the forests are *managed* by the wood products and pulp and paper industry. In such an arrangement, like other crops, trees are replanted wherever they are cut. In such circumstances, the industry can cut the trees in a way to allow easier reforestation. In Canada, the wood product and pulp and paper industry systematically replaces many of the trees it cuts, employing large numbers of summer workers for tree planting work.

In just 20 years, a teak plantation in Costa Rica can produce up to about 400 m of wood per hectare. As the natural teak forests of

Asia become more scarce or difficult to obtain, the prices commanded by plantation-grown teak grow higher every year. Other species such as mahogany grow slower than teak in Tropical America but are also extremely valuable. Faster growers include pine, eucalyptus, and *Gmelina*.

Reforestation, if several native species are used, can provide other benefits in addition to financial returns, including restoration of the soil, rejuvenation of local flora and fauna, and the capturing and sequestering of 38 tons of carbon dioxide per hectare per year.

The reestablishment of forests is not just simple tree planting. Forests are made up of a diversity of species and they build dead organic matter into soils over time. A major tree-planting program in a place like this would enhance the local climate and reduce the demands of burning large amounts of fossil fuels for cooling in the summer.

For Climate Change Mitigation

Forests absorb carbon dioxide through their photosynthesis cycle, and by using this idea, increasing forests with reforestation and discouraging deforestation will help mitigate global warming. Forest ecosystems are especially important to the global carbon cycle in two ways. First, they are responsible for moving around three billion tons of anthropogenic carbon every year. This amounts to about 30% of all carbon dioxide emissions from fossil fuels. Second, forest ecosystems are terrestrial carbon sinks in that they store large amounts of carbon which accounts for as much as double the amount of carbon in the atmosphere.

There are four major strategies available to mitigate carbon emissions through forestry activities: increase the amount of forested land through a reforestation process; increase the carbon density of existing forests at a stand and landscape scale; expand the use of forest products that will sustainably replace fossil-fuel emissions; and reduce carbon emissions that are caused from deforestation and degradation.

However, achieving the first strategy requires great effort of land transformation. For example, China has used 24 million ha of new forest plantation and natural forest regrowth to offset 21% of Chinese fossil fuel emissions in 2000. In theory, any tree would cover more forest area and absorb more carbon dioxide from the atmosphere. On the other hand, a genetically modified tree specimen might grow much faster than any other regular tree. Some of these trees are already being developed in the lumber and biofuel industries. These fast-growing trees would not only be planted for those industries but they can also be planted to help absorb carbon dioxide faster than regular trees. There are many projects around the world that use tree-planting as a way to offset carbon emissions, fighting climate change. For example, in China, Shanghai Roots & Shoots, a division of the Jane Goodall Institute launched The Million Tree Project in Kulun Qi, Inner Mongolia to plant one million trees to stop desertification and help curb climate change.

The rate of deforestation has huge potential toward a cost-effective contribution to protect the atmosphere's climate. At this point, there are 13 million ha of tropical regions that are deforested every year. These regions can reduce rates of deforestation by 50% by 2050.

A study from the National Centre for Atmospheric Research in Boulder, Colorado, USA, found that, unlike previous belief that forests in higher latitudes soak up a vast amount of carbon dioxide, more carbon dioxide is absorbed in tropical climates. Trees in temperate latitudes have a net warming effect on the climate. The heat that dark leaves absorb outweighs the carbon they soak up, therefore, tropical trees absorb carbon dioxide as well as being able to cool the planet by up to 0.7 °C.

Trees in tropical climates have, on average, larger, brighter, and more abundant leaves than non-tropical climates. The advantage of planting trees in a tropical setting is the quicker growth rate due to the longer rainy seasons. There is no need for the trees to hibernate and can therefore grow year-round.

An incredible portion of the Earth's biodiversity is situated in tropical areas. The lack of reforestation in tropical climates is putting a larger portion of species at risk of becoming endangered.

A study of the girth of 70,000 trees across Africa has shown that tropical forests are soaking up more carbon dioxide pollution than previously realised. The research suggests almost one fifth of fossil fuel emissions are absorbed by forests across Africa, Amazonia and Asia.

Simon Lewis, a climate expert at the University of Leeds, who led the study, said: "Tropical forest trees are absorbing about 18% of the carbon dioxide added to the atmosphere each year from burning fossil fuels, substantially buffering the rate of change."

Extensive forest resources placed anywhere in the world will not always have a positive impact. For example, large reforestation programs in boreal regions have a limited impact on climate mitigation. This is because it substitutes a bright snow-dominated region that reflects the sunlight with dark forest canopies.

On the other hand, a positive example would be reforestation projects in tropical regions, which would lead to a positive biophysical change such as the formation of clouds. These clouds would then reflect the sunlight, creating a positive impact on climate mitigation.

Incentives

Some incentives for reforestation can be as simple as a financial compensation. Streck and Scholz (2006) explain how a group of scientists from various institutions have developed a compensated reduction of deforestation approach which would reward developing countries that disrupt any further act of deforestation.

Countries that participate and take the option to reduce their emissions from deforestation during a committed period of time would receive financial compensation for the carbon dioxide emissions that they avoided. To raise the payments, the host country would issue government bonds or negotiate some kind of loan with a financial institution that would want to take part in the compensation promised to the other country.

The funds received by the country could be invested to help find alternatives to the extensive cutdown of forests. This whole process of cutting emissions would be voluntary, but once the country has agreed to lower their emissions they would be obligated to reduce their emissions. However, if a country was not able to meet their obligation, their target would get added to their next commitment period. The authors of these proposals see this as a solely government-to-government agreement; private entities would not participate in the compensation trades.

Examples

In Java, Indonesia each newlywed couple is to give whoever is sermonizing their wedding 5 seedlings to combat global warming. Each couple that wishes to have a divorce has to give 25 seedlings to whoever divorces them.

In Germany, reforestation is required as part of the federal forest law. 31% of Germany is forested, according to the second forest inventory of 2001–2003. The size of the forest area in Germany increased between the first and the second forest inventory due to forestation of degenerated bogs and agricultural areas.

Criticisms

Reforestation competes with other land uses such as food production, livestock grazing, and living space for further economic growth.

There is also the risk that through a forest fire or insect outbreak much of the stored carbon in a reforested area could make its way back to the atmosphere. Reduced harvesting rates and fire suppression have caused an increase in the forest biomass in the western United States over the past century. This causes an increase of about a factor of four in the frequency of fires due to longer and hotter dry seasons.

Revegetation

Revegetation is the process of replanting and rebuilding the soil of disturbed land. This may be a natural process produced by plant colonization and succession, or an artificial (manmade), accelerated process designed to repair damage to a landscape due to wildfire, mining, flood, or other cause. Originally the process was simply one of applying seed and fertilizer to disturbed lands, usually grasses or clover. The fibrous root network of grasses is useful for short-term erosion control, particularly on sloping ground. Establishing long-term plant communities requires the establishment of woody plants.

Soil Replacement

Mine reclamation may involve soil amendment, replacement, or creation, particularly for areas that have been strip mined or suffered severe erosion or soil compaction.

Mycorrhizal Communities

Mycorrhizae, symbiotic fungal-plant communities, are important to the success of revegetation efforts. Most woody plant species need these root-fungi communities to thrive, and nursery or greenhouse transplants may not have sufficient or correct mycorrhizae for good survival. Regional differences in ectomycorrhizal fungi may also affect the success of revegetation.

Tree Planting

Tree planting is the process of transplanting tree seedlings, generally for forestry, land reclamation, or landscaping purposes. It differs from the transplantation of larger trees in arboriculture, and from the lower cost but slower and less reliable distribution of tree seeds.

In silviculture the activity is known as reforestation, or afforestation, depending on whether the area being planted has or has not recently been forested. It involves planting seedlings over an area of land where the forest has been harvested or damaged by fire or disease or insects. Tree planting is carried out in many different parts of the world, and strategies may differ widely across nations and regions and among individual reforestation companies. Tree planting is grounded in forest science, and if performed properly can result in the successful regeneration of a deforested area. Reforestation is the commercial logging industry's answer to the large-scale destruction of old growth forests, but a planted forest rarely replicates the biodiversity and complexity of a natural forest.

Because trees remove carbon dioxide from the air as they grow, tree planting can be used as a geoengineering technique to remove CO_2 from the atmosphere.

By Country

Canada

Most tree planting in Canada is carried out by private reforestation companies. Reforestation companies compete with one another for contracts from logging companies, whose annual allowable cut for the following year is based upon how much money they invest into reforestation and other silvicultural practices. Treeplanting is typically piece work and tree prices can vary widely depending on the difficulty of the terrain and on the winning contract's bid price. As a result, there is a saying among planters: "There is no bad land, only bad contracts."

Tree planting crews often do not permanently reside in the areas where they work, thus much planting is based out of motels or bush camps. Bush camp accommodations usually consist of a mess tent, cook shack, dry goods tent, first aid tent, freshly dug outhouses, and a shower tent or trailer. Planters are responsible for bringing either a tent or car to sleep in. A camp also contains camp cooks and support staff.

Planting is carried out in accordance to the client's specifications, and planters are expected to learn the quality standards for each contract that they work on. Planted clearcuts are spot checked on a regular basis. Although quality concerns vary across contracts, spot checkers are typically looking for such things as: species appropriate site choice, species appropriate spacing, how tight the seedlings are in the ground, how straight the seedlings are, and whether or not the seedlings have been damaged. These concerns vary from region to region, and from contract to contract.

The average British Columbian planter plants 1 600 trees per day, but it is not uncommon for veterans to plant 2000-3000 trees per day while working in the interior. These numbers are higher in central and eastern Canada, where the terrain is generally faster, however the price per tree is slightly lower as a result. Average daily totals of 2500 are common, with experienced planters planting upwards of 5000 trees a day. Numbers as high as 7500 a day have been recorded. Planters typically work 8–10 hours per day with an additional 1 to 2 hours of (usually) unpaid travelling time. Work weeks on British Columbian planting contracts are usually 4–5 days long, with 1–2 days off. In Ontario, work weeks are generally 5–6 days long, with 1 day off.

Quite often, tree planting contractors will deduct some of the cost associated with the operation of the contract directly from the tree planter's daily earned wages. These imposed fees typically vary from $10 to $30 per day, and are referred to as "camp costs". In *some* cases, rookie tree planters end up owing their employer money for the first few pay periods.

Once inflation is factored in, real tree planter earnings have declined for many years in Canada. This has adversely affected the sector's ability to attract and retain workers. Higher wages and much better working conditions in many other industries, from construction, to oil and gas, and even information technology, has led to fewer Canadian young people wanting to plant trees.

Based on statistics for British Columbia, the average tree planter: lifts a cumulative weight of over 1,000 kilograms (2,200 lb), bends more than 200 times per hour, drives the shovel into the ground more than 200 times per hour and travels over 16 kilometres (9.9 mi) with a heavy load, every day of the entire season. The reforestation industry has an average annual injury rate of approximately 22 claims per 100 workers, per year. It is often difficult and sometimes dangerous.

Great Britain

Planting in Britain is commonly referred to as *restocking*, when it takes place on land that has recently been harvested. When occurring on previously unforested land it is known as *new planting*. Under the British system, in order to acquire the necessary permissions to clearcut, the landowner must agree a management plan with the

Forestry Commission (the regulatory body for all things forestry) which must include proposals for the re-establishment of tree cover on the land. Planting contractors will be engaged by the landowner/ management company, a contract drawn up and work will typically take place from November to April when the transplants are dormant.

Planting is part of the rotational nature of much British plantation forestry. Productive tree crops are planted and subsequently clearcut. Some form of soil cultivation may take place and the ground is then restocked. Where the production of timber is a management priority, a prescribed stocking density must be achieved. For coniferous species this will be a minimum of 2500 stems per hectare at year 5 (from planting). Planting at this density has been shown to favour the development of straighter knot-free logs.

Planters are normally paid under piece work terms and an experienced worker will plant around 1500 trees a day under most conditions.

Israel

With over 240 million trees planted, Israel is one of only two countries worldwide, that entered the 21st century with a net gain in the number of trees. Due to massive afforestation efforts, this fact echoed in diverse campaigns. The various forests of Israel today are mainly the result of a great afforestation campaign by the Jewish National Fund (JNF). Some well-known forests of contemporary Israel, out of the list of forests in Israel, are:

- the Hulda Forest or Herzl Forest (Ya'ar Hulda), a wood in central Israel between Kibbutz Hulda and Kibbutz Mishmar David, SE to Rehovot. Planted in 1907 by the Jewish National Fund, it was dedicated to the memory of Theodor Herzl and serves today as a public park and nature resort.
- the Forest of the Martyrs (Ya'ar HaKdoshim), a forest on the outskirts of Jerusalem, near Beit Meir. Planted as a memorial to the victims of the Holocaust, eventually it will contain six million trees as a symbol for the six million Jews who perished at the hands of the german Nazis in World War II.
- the Jerusalem Forest, a pine forest in the Judean Mountains west of Jerusalem. Planted during the 1950s by the Jewish National Fund and financed by private donors, the forest extends today over about 1.2 square kilometres, with the Yad Vashem museum near Mount Herzl.

- the Yatir Forest, located on the southern slopes of Mount Hebron, on the edge of the Negev Desert. Covering an area of 30,000 dunams (30 square kilometres), it is the largest planted forest in Israel. It is named after the ancient Levite city within its territory, Yatir, as written in the Torah: "And unto the children of Aaron the priest they gave Hebron with its suburbs, the city of refuge for the manslayer, and Libnah with its suburbs, and Jattir with its suburbs, and Eshtemoa with its suburbs" (Book of Joshua 21:13-14). Located within the Yatir forest now is an ancient Palestinian synagogue, the Anim synagogue (4th–7th centuries CE).

Afforesting the Desert

The state of Israel is even afforesting the Negev desert, which accounts for 60% of the country's land mass but remains sparsely populated. In 2006, the JNF signed a 49-year lease agreement with the State of Israel which gives it control over 30,000 hectares of Negev land for the development of forests.

The JNF's 600 million dollar Blueprint Negev aims to attract and build infrastructure for 250,000 new settlers in the Negev Desert, to meet the challenge of the massive Jewish immigration from the Soviet Union and Ethiopia. There is a multitude of modern israeli scientific research conducted in the Yair Forest to meet the challenge of climate change, which may result in rapid plant loss and desertification in certain circumstances. Studies of the Weizmann Institute of Science, in collaboration with the Desert Research Institute at Sde Boker, have shown that the trees function as a trap for carbon in the air. Shade provided by trees planted in the desert also reduces evaporation of the sparse rainfall.

The Yatir forest is a part of the NASA project FluxNet, a global network of micrometeorological tower sites used to measure the exchanges of carbon dioxide, water vapour, and energy between terrestrial ecosystem and atmosphere. The Arava Institute for Environmental Studies conducts research at Yatir forest that focuses on crops such as dates and grapes grown in the vicinity of Yatir forest. The research is part of a project aimed at introducing new crops into arid and saline zones.

Help for the Arab-Palestinian Authority

Since 2009, the JNF provided the Arab-Palestinian Authority with 3,000 tree seedlings for a forested area being developed on the edge of the new city of Rawabi, north of Ramallah.

Reforestation and Forest Wildfires

Approximately one thousand smaller forest fires are registered on average every year during the five fire-prone months. Half of them are caused by arson, hostile actions and arab or palestinian terrorists attacks. Ten thousand acres of hand-planted forest were destroyed by Katyusha rockets during the 2006 Lebanon War by Hezbollah militants at the Israeli northern border region. As a response, in summer 2006, JNF launched Operation Northern Renewal, a reforestation effort, which also replaced some topsoil that was burned away. The Mount Carmel forest fire, the largest forest fire in Israel's history, started on December 2, 2010 and burned 41 km^2 of forest.

New Zealand

Kaingaroa Forest in New Zealand is the largest planted forest in the southern hemisphere. It is one of the many plantation forests planted since European settlement. The Monterey Pine (*Pinus radiata*) is commonly used for plantations since a fast growing cultivar suitable for a wide range of conditions has been developed.

Government agencies, environmental organisations and private trusts carry out tree planting for conservation and climate change mitigation. While some work is carried out by private enterprise there are also planting days organised for volunteers. Landcare Research use planted forests for their EBEX21 system for greenhouse gas emissions mitigations.

United States

Trees for the Future is a non-profit organisation that plants trees in developing countries to improve land management.

Role in Climate Change

The development of markets for tradeable pollution permits in recent years have opened up a new source of funding for tree planting projects: carbon offsets. The creation of carbon offsets from tree planting projects hinges on the notion that trees help to mitigate climate change by sequestering carbon dioxide as they grow. However, the science linking trees and climate change is largely unsettled, and trees remain a controversial source of offsets.

Climate Impacts

Climate scientists believe that human-induced global deforestation is responsible for 18-25% of global climate change. The United Nations, World Bank and other leading nongovernmental organisations are encouraging reforestation, avoided deforestation and other projects that encourage tree planting to mitigate the effects of climate change.

Trees sequester carbon through photosynthesis, converting carbon dioxide and water into molecular dioxygen (O_2) and plant organic matter, such as carbohydrates (e.g., cellulose). Hence, forests that grow in area or density and thus increase in organic biomass will reduce atmospheric CO_2 levels. (Carbon is released as CO_2 if a tree or its lumber burns or decays, but as long as the forest is able to grow back at the same rate as its biomass is lost due to oxidation of organic carbon, the net result is carbon neutral.)

In their 2001 assessment, the IPCC estimated the potential of biological mitigation options (mainly tree planting) is on the order of 100 Gigatonnes of carbon (cumulative) by 2050, equivalent to about 10% to 20% of projected fossil fuel emissions during that period.

However, the global cooling effect of forests from carbon sequestration is not the only factor to be considered. For example, the planting of new forests may initially release some of the area's existing carbon stores into the atmosphere. Specifically, the conversion of peat bogs into oil palm plantations has made Indonesia the world's third largest producer of greenhouse gases.

Compared to less vegetated lands, forests affect climate in three main ways:

- Cooling the Earth by functioning as carbon sinks, and adding water vapour to the atmosphere and thereby increasing cloudiness.
- Warming the Earth by absorbing a high percentage of sunlight due to the low reflectivity of a forest's dark surfaces. This warming effect, or reduced albedo, is large where evergreen forests, which have very low reflectivity, shade snow cover, which is highly reflective.

To date, most tree planting offsets strategies have taken only the first effect into account. A study published in December 2005 combined all these effects and found that tropical forestation has a large net cooling effect, because of increased cloudiness and because of high tropical growth and carbon sequestration rates.

Trees grow three times faster in the tropics than in temperate zones; each tree in the rainy tropics removes about 22 kilograms (50 pounds) of carbon dioxide from the atmosphere each year. However, this study found little to no net global cooling from tree planting in temperate climates, where warming due to sunlight absorption by trees counteracts the global cooling effect of carbon sequestration. Furthermore, this study confirmed earlier findings that reforestation of colder regions — where long periods of snow cover, evergreen trees, and slow sequestration rates prevail — probably results in global warming. According to Ken Caldeira, a study co-author from the Carnegie Institution for Science, "To plant forests outside of the tropics to mitigate climate change is a waste of time.".

His premise that grassland reflects more sun, keeping temperatures lower, is, however, applicable only in arid regions. A well-watered lawn, for example, is as green as a tree, but absorbs far less CO_2. Deciduous trees also have the advantage of providing shade in the summer and sunlight in the winter; so these trees, when planted close to houses, can be utilised to help increase energy efficiency of these houses.

This study remains controversial and criticized for assuming dark coloured trees might replace the frozen, white tundra in the upper northern hemisphere. Regular tree planting projects typically take place on lands that are only slightly different in colour. The warming impact was also measured over hundreds of years, rather than a 30-70 year time horizon most climate experts believe we have to fix climate change.

Furthermore, the described warming effect (of temperate and boreal latitude forest) is only apparent once the trees have grown to create a dense 'close canopy', and it is at precisely this point that trees grown for offset purposes should be harvested and their absorbed carbon fixed for the long-term as timber.

Costs

While the benefits of tree planting are subject to debate, the costs are low compared to many other mitigation options. The IPCC has concluded that "The mitigation costs through forestry can be quite modest (US$0.1–US$20 / metric ton carbon dioxide) in some tropical developing countries.

The costs of biological mitigation, therefore, are low compared to those of many other alternative measures". The cost effectiveness of tropical reforestation is due not only to growth rate, but also to farmers from tropical developing countries who voluntarily plant and nurture tree species which can improve the productivity of their lands. As little as US$90 will plant 900 trees, enough to annually remove as much carbon dioxide as is annually generated by the fossil-fuel usage of an average United States resident.

Types of Trees Planted

The type of tree planted may have great influence on the environmental outcomes. Planting the wrong kind of trees, such as monocultures of eucalyptus where they are not native species, can devastate the lands of the local people. However, it is often much more profitable to outside interests to plant non-native fast-growing trees, such as eucalyptus or pine (e.g., *Pinus radiata* or *Pinus caribaea*), even though the environmental and biodiversity benefits of such monoculture plantations are not comparable to native forest, and such offset projects are frequently objects of controversy.

To promote the growth of native ecosystems, many environmentalists advocate only indigenous trees be planted. A practical solution is to plant tough, fast-growing native tree species which begin rebuilding the land. Planting non-invasive trees that assist in the natural return of indigenous species is called "assisted natural regeneration." There are many such species that can be planted, of which about 12 are in widespread use, such as *Leucaena leucocephala*.

A skyline is the overall or partial view of a city's tall buildings and structures consisting of many skyscrapers in front of the sky in the background. It can also be described as the artificial horizon that a city's overall structure creates. Skylines serve as a kind of fingerprint of a city, as no two skylines are alike. For this reason news and sports programs, television shows, and movies often display the skyline of a city to set location. *The Sky Line of New York City* was a new term in 1896, when it was the title of a colour lithograph by Charles Graham for the colour supplement of the *New-York Journal*.

In general, larger cities have broader and taller skylines, though lower density cities often have smaller skylines than expected for city size. Taller buildings are found where either land value or desire for visibility is higher, and the tallest buildings in a city are usually office buildings. Because of this, the skyline of a city can be seen as symbolic of the city's influence and economy.

Cityscape

A cityscape is the urban equivalent of a landscape. *Townscape* is roughly synonymous with *cityscape,* though it implies the same difference in urban size and density (and even modernity) implicit in the difference between the words *city* and *town*. In urban design the terms refer to the configuration of built forms and interstitial space. In the visual arts a cityscape (urban landscape) is an artistic representation, such as a painting, drawing, print or photograph, of the physical aspects of a city or urban area.

History of Cityscapes in Art

From the first century A.D. dates a fresco at the Baths of Trajan in Rome depicting a bird's eye view of an ancient city. In the Middle Ages cityscapes appeared as a background for portraits and biblical themes. From the 16th up to the 18th century numerous copperplate prints and etchings were made showing cities in bird's eye view. The function of these prints was to provide a map-like overview.

Halfway the 17th century the cityscape became an independent genre in the Netherlands. In his famous 'View of Delft' in 1660-1661 Jan Vermeer painted a quite accurate portrait of the city Delft. Cities like Amsterdam, Haarlem and The Hague also became popular subjects for paintings. Painters from other European countries (i.e. Great Britain, France, Germany) followed the Dutch example. The 18th century was a flourishing period for cityscape painting in Venice (Canaletto, Guardi).

At the end of the 19th century the impressionists focused on the atmosphere and dynamics of everyday life in the city. Suburban and industrial areas, building sites and railway yards also became subjects for cityscapes. During the 20th century attention became focused on abstract and conceptual art, and thus the production of cityscapes declined. American painter Edward Hopper, who stayed loyal to figurative painting, created intriguing images of the American scene. With a revival of figurative art at the end of the 20th century comes a revaluation of the cityscape. Well-known living cityscape painters are Rackstraw Downes, Antonio López García, and Richard Estes. American artist Yvonne Jacquette has made a speciality of aerial cityscapes. Stephen Wiltshire, a London born artist with autism, is known for his panoramic cityscape renderings composed from memory, usually after taking a short overhead view of the city he is about to draw.

Another London born artist, Brian Whelan uses multi-perspectives and contradictory scales, bending, twisting and organising the urban chaos into harmony.

Cityscape Painters

- Johann Berthelsen
- George Hendrik Breitner
- Gustave Caillebotte

- Canaletto
- Edouard Leon Cortés
- John Atkinson Grimshaw
- Childe Hassam
- Jan van der Heyden
- Isaac Israëls
- Matthäus Merian
- Camille Pissarro
- Paul Signac
- Alfred Sisley
- Jan Vermeer
- Brian Whelan
- James McNeill Whistler
- Guy C. Wiggins
- Stephen Wiltshire.

Bibliography

Badiru, A.: *Handbook of Industrial and Systems Engineering*, CRC Press, Delhi, 2005.

Beaty H. Wayne: *Standard Handbook for Electrical Engineers*. New York, NY: McGraw-Hill, 2000.

Beizer, Boris: *Software Testing Techniques,* International Thomson Computer Press, Boston, MA., 2004.

Benjamin Henry: *The Engineering Drawings of Benjamin Henry Latrobe,* Yale University Press, NY, 1980.

Bennett, Hugh Hammond : *Elements of Soil Conservation*, Biotech Books, Delhi, 2009.

Bensted, J.: *Structure and Performance of Cements*, Spon Press, New York, 2001.

Bernard S. Matisoff: *Handbook of Electronics Manufacturing Engineering*, Chapman & Hall, New York, 1997.

Bhakar S.R. : *Ground Water Hydrology : Theory and Practice*, Agrotech, Delhi, 2009.

Booker, P. J.: *A History of Engineering Drawing*, Chatto & Windus, London, 1963.

Branan, C.R.: *Rules of Thumb for Chemical Engineers*, Gulf Professional Publishing, 2005.

Burnell, George: *Rudimentary Treatise on Limes, Cements, Mortars, Concretes, Mastics, Plastering,* Virtue and Co., London, 1870.

Burstall, A. F.: *A History of Mechanical Engineering,* Faber and Faber, London, 1963.

Carl Zenz: *Occupational Medicine: Principles and Practical Applications*. Chicago, Year Book Medical Publishers, 1988.

Chandra, Ramesh and Satish Kumar Singh: *Fundamentals and Management of Soil Quality*, Westville Pub, Delhi, 2009.

Chatterjee, A. K. : *Advances in Cement Technology,* Pergamon Press, N.Y., 1983.

China, T.: *Building Industry, Pride of the Pacific,* Honolulu, Hawaii, 1991.

Clafin, A. A.: *Technology and Industrial Efficiency: Proceedings of the Congress of Technology 1911*, McGraw Hill, New York, 1911.

Colton, C. K.: *Perspectives in Chemical Engineering*, Academic, Boston, 1991.

Conaway, C.F.: *Petroleum Industry: A Nontechnical Guide*, Pennwell Publishing, Tulsa, USA, 1999.

Craig, Ed: *Gas Metal Arc & Flux Cored Welding Parameters*, Weldtrain, Chicago, 1991.

Dasgupta, S.: *Technology and Creativity*, Oxford University Press, New York, 1996.

Davey, N.: *History of Building Materials*, Phoenix House, London, 1961.

David Cebon: *Materials: Engineering, Science, Processing and Design*, Butterworth-Heinemann, UK, 2007.

Davis, E. H.: *The Profession of a Civil Engineer*, Sydney University Press, Sydney, 1979.

Dimes, Francis B.: *Conservation of Building and Decorative Stone*, Heineman, Butterworth, 1990.

Domminghaus, H.: *Plastics for Engineers: Materials, Properties, Applications*, Hanser Publishers, Munich, 1993.

Duncan, G. A.: *Estimating Greenhouse Ventilation Requirements*, Dept. Agric. Engr., Univ. Kentucky, Lexington, 1973.

Eveleigh, E.W.: *Introduction to Control Systems Design*, McGraw-Hill, 1972.

Fabrycky, W.: *Systems Engineering and Analysis*, Prentice-Hall, NY, 2005.

Felder, R.M. *Elementary Principles of Chemical Processes*, John Wiley and Sons, New York, 2004.

Finch, J. K.: *Engineering Classics*, Kensington, Cedar Press, MD, 1978.

Fitchen, J.: *Building Construction Before Mechanization*, MIT Press, London, 1988.

Francis B.: *Conservation of Building and Decorative Stone*, Heineman, Butterworth, 1990.

Fred, Aftalion: *A History of the International Chemical Industry*, University of Pennsylvania Press, Pennsylvania, 1991.

Gibson, Dan: *The Nabataeans, Builders of Petra*, CanBooks, Amman, Jordan, 2002.

Gille, B.: *Engineers of the Renaissance*, MIT Press, Cambridge, 1966.

Glazer, A.M.: *Space Groups for Scientists and Engineers*, Academic Press, Boston, 1990.

Groover, Mikell P.: *Fundamentals of Modern Manufacturing*, John Wiley & Sons, INC, New York, 2002.

Haber, L. F. *The Chemical Industry: 1900-1930, International Growth and Technological Change*. Oxford: Claredon Press, 1971.

Hasan, Syed Danish : *Civil Engineering Materials and Their Testing*, Narosa, Delhi, 2011.

Hill, D.: *A History of Engineering in Classical and Medieval Times*, Open Court, La Salle, IL, 1984.

Houghton, J.T.: *The Physics of Atmospheres*, Cambridge University Press, UK, 1998.

Hyne, N.J.: *Nontechnical Guide to Petroleum Geology, Exploration, Drilling and Production*, Pennwell Publishing, Tulsa, USA, 2001.

James Brown: *Advanced Machining Technology Handbook,* McGraw-Hill, New York, 1998.

Jenkin, C. F.: *An Inaugural Lecture of the Training of the Engineering Profession,* University of Oxford Press, 1908.

John Hays. *The Chemical Engineer, The Profession of Engineering.* New York: John Wiley & Sons, 1929.

Johnston, Jr., E.R.: *Mechanics of Materials*, McGraw-Hill, Inc., New York, 1981.

Joseph A. Brink Jr.: *The Chemical Process Industries*, McGraw Hill, New York, 1977.

Joshi, Jagat Pati : *Harappan Architecture And Civil Engineering*, Rupa Pub, Delhi, 2008.

Kalpakjian, Serope: *Manufacturing Engineering and Technology*, Prentice Hall, 2001.

Kennedy, M. L. : *The Technology of Cement and Concrete,* John Wiley & Sons, New York, 1955.

Kunda, Gideon: *Engineering Culture: Control and Commitment in a High-Tech Corporation.* Philadelphia: Temple University Press, 1992.

Lea, F. M.: *The Chemistry of Cement and Concrete,* St. Martin Press, New York, 1956.

Leavitt, H.J.: *Applied Organizational Change in Industry: Structural, Technological and Humanistic Approaches*, Rand McNally, Chicago, 1965.

Lewis, P.H. and Yang, C.: *Basic Control Systems Engineering*, Prentice Hall, 1997.

MacCullough, G.H.: *Elements of Strength of Materials,* Van Nostrand Co., New York, 1957.

Malinowski, R.: *Concrete and Mortars in Ancient Aqueducts,* Concrete International: Design and Construction. Detroit, Michigan, 1979.

Mason, J.: *Commercial Hydroponics*, Kangaroo Press, Kent., NSW, Australia, 1990.

McCormick, T. : *Metal Technics*, Brooks Waterloo-Jacaranda Press, Milton, Qld, 1989.

Meyn, S.P. : *Control Techniques for Complex Networks*, Cambridge University Press, New York, NY, 2008

Mielenz, R. C.: *Concrete as a Modern Material, Modern Materials,* Academic Press, New York, 1965.

Nicholas P.: *Handbook of Chemical Engineering Calculations*, McGraw-Hill, New York, 2004.

Orchard, D. F.: *Concrete Technology,* Applied Science Publishers, London, 1973.

Parkash, S.: *Refining Processes Handbook*, Elsevier, Amsterdam, 2003.

Patel, Sheelwant: *Indian Forests : Soil Water and Bio-Environment Conservation*, Pointer, Delhi, 2005.

Patton, W. M.: *A Treatise on Civil Engineering,* John Wiley and Son, N.Y., 1905.

Peckworth, H. F.: *Concrete Pipe Handbook,* American Concrete Pipe Assoc., Chicago, 1959.

Popovics, S.: *Concrete-Making Materials,* McGraw-Hill Book Co., New York, 1979.

Richards, E. H.: *Technology and Industrial Efficiency: Proceedings of the Congress of Technology 1911*: McGraw Hill, New York, 1911.

Robert S.: *Chemistry and Technology of Lime and Limestone,* John Wiley and Sons, NY, 1966.

Smith, J.M.; H.C.van Ness, M.Abbott: *Introduction to Chemical Engineering Thermodynamics*, McGraw-Hill Education, New York, 2005.

Stine, Jeffrey K. : *Mixing the Waters: Environment, Politics, and the Building of the Tennessee-Tombigbee Waterway*. University of Akron Press, 1993.

Taylor, W. H.: *Concrete Technology and Practice,* Angus and Robertson Publisher, 1965.

Turner, W.: *Introduction to Industrial and Systems Engineering*, Prentice Hall, NY, 1992.

Utkin, V.I.: *Sliding Modes in Control and Optimization*, Springer-Verlag, Berlin, 1992.

Waddell, J. J.: *Construction Materials Ready Reference Manual,* McGrawHill Book Co., New York, 1985.

Weiler, T.C. and M. Sailus: *Water and Nutrient Management for Greenhouses*, Northeast Regional Agricultural Engineering Service Cooperative Extension. Ithaca, New York, USA, 1996.

Wilde, S.A. : *Forest Soils and Forest Growth*, Periodicals, Delhi, 1991.

William Alec: *The Chemical Industry*. London: Longman Group Limited, 1971.

William D.: *Materials Science and Engineering – An Introduction*, John Wiley and Sons, UK, 2000.

Index

T

W

Z

❑❑❑